Making Sense of Statistics

A Conceptual Overview

Second Edition

Fred Pyrczak

California State University, Los Angeles

 P y r c z a k   P u b l i s h i n g

P.O. Box 39731 • Los Angeles, CA 90039

Editorial assistance provided by Brenda Koplin, Sharon Young, Elaine Fuess, and Cheryl Alcorn.

Cover design by Robert Kibler and Larry Nichols.

Printed in the United States of America.

ISBN 1-884585-33-7

Contents

Introduction   *v*

Part A   The Research Context

1. The Empirical Approach to Knowledge   1

2. Types of Empirical Research   5

3. Introduction to Sampling   9

4. Scales of Measurement   15

5. Descriptive and Inferential Statistics   19

Part B   Descriptive Statistics

6. Frequencies, Percentages, and
   Proportions   23

7. Shapes of Distributions   27

8. The Mean: An Average   33

9. Mean, Median, and Mode   37

10. Range and Interquartile Range   41

11. Standard Deviation   45

12. Correlation   49

13. Pearson *r*   55

14. Coefficient of Determination   59

Part C   Inferential Statistics

15. Sampling Concepts   63

16. Standard Error of the Mean   69

17. Introduction to the Null Hypothesis   75

18. Decisions About the Null Hypothesis   81

19. Introduction to the *t* Test   87

20. Reporting the Results of *t* Tests   93

21. One-Way ANOVA   97

22. Two-Way ANOVA   103

23. Chi Square   113

Continued

Appendices

A. Computation of the Standard
   Deviation   117

B. Notes on Interpreting Pearson *r* and
   Linear Regression   119

C. Table of Random Numbers   121

D. More About the Null Hypothesis   122

Comprehensive Review Questions   123

iv

Introduction

Four types of students need a conceptual overview of statistics.

(1) Those who are preparing to be consumers of empirical research and need basic concepts in order to interpret the statistics reported by others in journals, at conferences, and in reports prepared by their supervisors and co-workers.

(2) Those who are taking a traditional introductory statistics course and are getting lost in the details of deriving formulas, computing statistics, and sorting through an overload of theory. For them, this book is an ideal supplement.

(3) Those who are taking an applied introductory statistics course in which computers are used for all major calculations but need a text that helps them understand the meaning of their output. For them, this book may be used as the main textbook.

(4) Those who have taken a statistics course but need to review essential concepts in preparation for a subsequent course for which statistics is a prerequisite or in preparation for planning their theses and dissertations.

Coverage

The coverage here is highly selective. To be included in this book, a statistical technique had to be one that is almost universally covered in introductory statistics courses and is widely reported in journals. Once students understand and feel comfortable with the statistics that meet these criteria, they should find it easy to master additional statistical concepts.

Computations and Formulas

You will find no formulas and very few computational procedures described in the body of this book. I have assumed that for some students a noncomputational approach is best for developing an understanding of the meaning of statistics. For students who are becoming consumers of research, this may be all that is needed. For others, computers can handle the computations; such students need to know which statistics are available for various purposes and how to interpret them.

Students who are mathematically inclined may find that computations help them understand statistical concepts; they may wish to supplement this book with others described below.

Steps in Using This Book

To get the most from this book, I suggest that you follow these steps: (1) read a section while ignoring all references to footnotes and appendices since these contain information that may distract you as you master essential concepts; (2) read the section again, pausing to read the footnotes and appendices when you encounter references to them; (3) read the summary statements in the sidebars for review; and (4) answer the exercise questions at the end of the section.

If you have not studied statistics, you will find that most of the concepts in this book are entirely new to you. It is not realistic to expect to skim the text once to achieve full mastery. In addition, pilot tests indicate that students who read each section twice before answering the end-of-section questions mastered the concepts more quickly than those who read it only once. The latter did more hunting and picking through the text to look for answers, which is a slow process that does not provide an overview and leads to errors. Thus, you should save time and achieve greater mastery by following the recommended steps faithfully.

New to the Second Edition

The major change from the First to Second Editions is the addition of a section at the end of this book titled Comprehensive Review Questions. In it, you will find five multiple-choice and true-false questions on each of the 23 sections of this book. While the questions may be assigned at any point in a course, they are intended especially for students who need to review the contents of this book in preparation for a final examination. Students who are using the review questions for this purpose should keep in mind that the questions do *not* cover every point an instructor might cover on an examination. Hence, the body of the book, as well as notes taken during the course, should also be reviewed in preparation for an examination.

Acknowledgments

I am grateful to Robert Morman and Patricia Bates Simun, both of California State University, Los Angeles, and to Roger A. Stewart, University of Wyoming, for their helpful comments on the First Edition of this book. Errors and omissions remain the responsibility of the author.

Fred Pyrczak
Los Angeles, California

Notes:

Section 1

The Empirical Approach
to Knowledge

In the *empirical approach* to knowledge, we observe events to obtain information.[1] This approach is used when we make *everyday observations*. For example, if we observe that a traffic officer regularly hides behind the bushes at a certain intersection and frequently issues tickets at that location, we might say we *know* that someone who runs a stop sign at that intersection is likely to get a ticket. Unfortunately, generalizations based on everyday observations are often misleading. Here's an example:

> Suppose you observe that most of your friends and acquaintances plan to vote in favor of a school bond measure to build new schools. Unless they are a good cross section of the electorate, which is unlikely, you may be wrong if you generalize to the population of voters and predict that the measure will pass in the election.

Most scientists also use the empirical approach to acquire knowledge. When they do, we say that they are engaging in *empirical research*. A major distinction between empirical research and everyday observation is that empirical research is planned in advance of making observations. Based upon a theory or hunch, scientists develop research questions and then plan *whom*, *how*, and *when* to observe in order to answer them:[2]

(1) They plan *whom* (or *what*) to observe. It could be all mentally ill patients in a hospital ward or all sociology textbooks used in a

The empirical approach is based on observation.

Everyday observation is an example of the empirical approach.

Generalizations based on everyday observations often are misleading.

Scientists plan *whom*, *how*, and *when* to observe.

---

[1] Examples of other approaches are (1) *deduction* as when we deduce a proof in mathematics based on certain assumptions and definitions and (2) *reliance on authority* such as relying on a dictator's pronouncements as a source of knowledge.

[2] Research questions are the heart of all empirical research. When a scientist predicts the answer to a research question, we say that she has a *hypothesis*.

school district. This is known as the *population*. When a population is large, scientists often plan to observe only a *sample*. Planning how to draw an adequate sample is, of course, important in conducting valid research.[3]

Planning how to draw a *sample* from a *population* is important in conducting valid research.

(2) They plan *how* to observe by deciding whether to construct new measuring instruments or select instruments that have been developed by others. For example, they might review existing multiple-choice tests and select those that are most valid for answering their research questions. They might also modify existing interview schedules, questionnaires, personality scales, etc., to make them more suitable for their research. Finally, they might build entirely new instruments from scratch.

Measuring instruments are constructed or selected.

(3) They plan *when* the observations will be made and *under what circumstances*. For example, will the observations be made in the morning in a quiet room or in a shopping mall on a busy Saturday afternoon? Scientists realize that the timing and circumstances of their observations may affect the results of their investigations.[4]

Timing and circumstances of the observations may affect the results.

Unfortunately, not all plans are good, and even the best plans often cannot be fully executed because of physical, ethical, legal, and financial constraints. Thus, empirical research varies in quality, and flawed research can be just as misleading as everyday observations often are.

Flawed research can be misleading.

The observations that scientists make result in data. The data might be the names of political candidates for whom subjects plan to vote or they might be scores on a scale that measures depression. Large amounts of data need to be organized and summarized, which is a primary function of statistical analysis.[5] For example, we could summarize the data from an election poll by computing the percentage of subjects who plan to vote for each candidate, or we could summarize depression scores by computing an average score. These and many other statistics are described in this book.

Large amounts of data need to be organized and summarized, which is a function of statistics.

---

[3]Sampling methods are described in Sections 3 and 15.

[4]In addition, these factors may affect the generalizability of their findings. For example, people's behavior in a busy mall may be quite different from their behavior in other settings. Thus, the findings might apply only to the circumstances under which the observations were made.

[5]Planning in advance how the data will be analyzed statistically is also a characteristic of good research.

Exercise for Section 1

Factual Questions

1. The empirical approach to knowledge is primarily based on what?

2. Does everyday observation employ the empirical approach?

3. When scientists use the empirical approach and plan its use in advance, we say that they are engaging in what?

4. Which type of planning involves identifying a sample and deciding whether to sample? (Circle one.)
    A. whom   B. how   C. when

5. Which type of planning involves constructing or selecting measuring instruments? (Circle one.)
    A. whom   B. how   C. when

6. A sample is a subgroup of what?

7. What is a primary function of statistical analysis?

Questions for Discussion

8. Do you think that the opinions of your friends and acquaintances are good predictors of the outcomes of elections? Why? Why not?

9. Try to remember an instance in which you were misled by everyday observation and briefly describe it here.

10. Try to remember an instance in which you read or heard about empirical research that you suspected was flawed. Briefly describe here what you suspected.

Section 2

Types of Empirical Research

A fundamental distinction is whether research is *experimental* or *descriptive*. An *experiment* is a study in which treatments are given to see how the subjects respond to them. We all conduct informal experiments. Here are some examples:

> We might try a new laundry detergent (the treatment) to see if our clothes are cleaner (the response) than when we used our old brand.

> A teacher might bring to class a new short story (the treatment) to see if students enjoy it (the response).

> A waiter might try being more friendly (the treatment) to see if it increases his tips (the response).

In an experiment, a set of treatments is called the *independent variable*, and the responses are called the *dependent variable*. Independent variables are administered so that we can observe possible changes in dependent variables.

Clearly, the purpose of experiments is to identify *cause-and-effect relationships*. When our hypothetical waiter tries being more friendly, he is interested in finding out whether the increased friendliness *causes* increased tips.

Unfortunately, informal experiments can be misleading. For example, suppose the waiter notices that when he is more friendly, he gets larger tips than when he was less friendly. Did the increased friendliness *cause* the increase in the tips? The answer is not clear. Perhaps, by chance, the evening he tried being more friendly, he happened to have customers who were more generous. Perhaps Dear Abby published a column that day on tipping, which urged people to treat waiters and

---

An *experiment* is a study in which treatments are given and responses to them observed.

Treatments are called the *independent variable* and responses are called the *dependent variable*.

The purpose of experiments is to determine *cause-and-effect relationships*.

waitresses more fairly. Perhaps the waiter was not only more friendly but, unconsciously, also more efficient, and the increased efficiency and not the increased friendliness caused the increase in tips. The possible alternative explanations are almost endless unless the experiment is planned in advance to eliminate them.

Unless it is properly planned, there may be many alternative explanations for the results of an experiment.

How could we conduct a formal experiment on friendliness and tipping that would have a clearer interpretation? Simply by having an appropriate control condition. For example, we could have the waiter be more friendly to every other party on which he waits. Those who receive the normal amount of friendliness would be referred to as the *control group*. In addition, we could monitor the waiter's behavior to be sure that it was the same in all respects for both groups of customers except for how friendly he was. We could then use statistics to compare the average tips earned under the more friendly condition with those earned under the less friendly condition.[1]

An appropriate control condition is an essential characteristic of good experiments.

A *descriptive study* is one in which we observe only to determine the status of what exists at a given point in time; we do *not* administer treatments. A good example is a survey in which we wish to determine subjects' attitudes toward something. In such a study, scientists strive *not* to change the subjects' attitudes, for example, by avoiding leading questions and having the interviewers be neutral in tone and mannerisms. If the sample is properly drawn and well-crafted questions are asked under the right circumstances, scientists can describe the subjects' attitudes by computing and reporting appropriate statistics. They cannot, however, properly discuss how to change attitudes on the basis of this type of study—to do this, they would need to conduct an experiment.[2]

A *descriptive study* is one in which we observe only to determine what exists.

---

[1]Strictly speaking, the results would apply only to this waiter working in this restaurant on nights similar to the ones on which we conducted the experiment. In order to generalize to other waiters and waitresses in other restaurants, we would need to include a sample of them in the experiment.

[2]For some important causal questions, it is not possible to conduct an experiment. For example, in studying the effects of smoking and health, it would be unethical to force some human subjects to smoke while forbidding others to do so. In such situations, scientists must use descriptive data when discussing causality.

Exercise for Section 2

Factual Questions

1. In which type of study are treatments given in order to see how subjects respond?

2. In an experiment, are the responses the *independent variable* or the *dependent variable*?

3. What is the purpose of an experiment?

4. What is an essential characteristic of good experiments?

5. In which type of study do scientists strive not to change the subjects?

6. What is the definition of a *descriptive study*?

Questions for Discussion

7. Briefly describe an informal experiment that you recently conducted. Were there alternative explanations for the responses you observed?

8. Do you think that both *descriptive* and *experimental* studies have a legitimate role to play in the acquisition of scientific knowledge? Why? Why not?

Notes:

Section 3

Introduction to Sampling

A *population* consists of all members of a group in which a scientist has an interest.[1] It may be small such as all psychiatrists who are affiliated with a particular hospital, or it may be large such as all high school seniors in a state. When populations are large, scientists usually sample. A *sample* is a subgroup of a population. For example, we might be interested in the attitudes of all registered nurses in Texas toward people with AIDS. The nurses would constitute our population. If we administered an AIDS attitude scale to all of these nurses, we would be studying the population and the summarized results (such as averages) would be referred to as *parameters*. If we studied only a sample of the nurses, the summarized results would be referred to as *statistics*.

No matter how we sample, it is always possible that the *statistics* we obtain do not accurately reflect the *population parameters* that we would have obtained if we had studied the entire population. In fact, we always expect some amount of error when we have sampled.

If sampling creates errors, why do we sample? First, it is not always possible for economic and physical reasons to examine an entire population. Second, with proper sampling, we can obtain highly reliable results for which we can estimate the amount of error to allow for in our interpretations of the data.

The most important characteristic of a good sample is that it is free of *bias*. A bias exists whenever some members of a population have a greater chance of being selected for inclusion in the sample than others. Here are some examples of biased samples:

A professor wishes to study the attitudes of all sophomores at a college (the population) but asks only those enrolled in her introductory psychology class (the sample) to participate in the

A *population* consists of all members of a group.

A *sample* is a subgroup of a population.

Populations yield *parameters*.

Samples yield *statistics*.

A *bias* is created when some members of a population have a greater chance of being selected than others.

---

[1] All members of a population have at least one characteristic in common.

study. Note that only those in the class have a chance of being selected; other sophomores have no chance.

A person wishes to predict the results of a citywide election (the population) but asks the intentions only of voters he encounters in a large shopping mall (the sample). Note that only those he decides to approach in the mall have a chance of being selected; other voters have no chance.

A magazine editor wishes to determine the opinions of all rifle owners (the population) on a gun-control measure but mails questionnaires only to those who subscribe to her magazine, which appeals to sporting enthusiasts (the sample). Note that only subscribers have a chance to respond; other rifle owners have no chance.

In the examples above, *samples of convenience* (or *accidental samples*) were used, increasing the odds that some will be selected and reducing the odds that others will—but there is an additional problem. Even those who have a chance of being included may refuse to participate. This problem is often referred to as *volunteerism*. Volunteerism is presumed to create a bias because those who decide not to participate have no chance of being included. Furthermore, many studies comparing participants with nonparticipants suggest that participants tend to be more highly educated and from higher socioeconomic status (SES) groups than their counterparts. Efforts to reduce the effects of volunteerism include offering rewards, stressing to potential participants the importance of the study, and making it easy for people to respond such as providing them with a stamped, self-addressed envelope.

Samples of convenience are biased.

Volunteerism in sampling is presumed to create a bias.

To eliminate bias, some form of *random sampling* is needed. A classic form of random sampling is *simple random sampling*.[2] This technique gives each member of a population an equal chance of being selected. A simple way to accomplish this with a small population is to put the names of all members of a population on slips of paper, thoroughly mix the slips, and have a blindfolded assistant select the number desired

In *simple random sampling* each member of a population is given an equal chance of being selected.

---

[2]Another method for selecting a *simple random sample* and other types of random samples are described in Section 15.

for the sample. After the names have been selected, efforts must be made to encourage participation of all those selected. If some refuse, as often happens, we have a biased sample even though we started by giving everyone an equal chance to have his or her name selected.

But let us suppose that we are fortunate. We have selected names using simple random sampling, and we have obtained the cooperation of everyone selected. In this case, we have an *unbiased sample*. Can we be certain that the results we obtain from the sample accurately reflect those we would have obtained by studying the population? Certainly not! We now have the possibility of random errors, which are simply called *sampling errors* by statisticians. At random (i.e., by chance) we may have selected a disproportionately large number of males, Democrats, low SES group members, etc. Such errors make the sample unrepresentative and, therefore, may lead to incorrect results.

If both biased and unbiased sampling are subject to error, why do we prefer unbiased random sampling? We prefer it for two reasons: (1) inferential statistics, which are described in Part C of this book, allow us to estimate the amount of error to allow for when analyzing the results from unbiased samples, and (2) the amount of error obtained from unbiased samples is small when large samples are used.

It is important to note that selecting a very large biased sample does not reduce potential errors. For example, if the person who is trying to predict the results of a citywide election in the earlier example is very persistent and spends weeks at the shopping mall asking all registered voters that he encounters how they intend to vote, he will obtain a very large sample of people who may differ from the population of voters in various ways — such as being more affluent, having more time to spend shopping, etc. Increasing the sample size, in this case, will not reduce the amount of error due to bias. Considerations in determining an appropriate sample size are discussed in Section 15.

I do not mean to leave you with the impression that all research in which biased samples are used is worthless. There are many situations in which scientists have no choice but to use biased samples. For example, for ethical and legal reasons, much medical research is done on volunteers who are willing to take the risk of using a new medication or undergoing a new procedure. If promising results are obtained in initial

Simple random sampling identifies an *unbiased sample*.

Random sampling produces *sampling errors*.

Selecting a very large sample does not correct for errors due to bias.

Often scientists have no choice but to use biased samples.

studies, larger studies with better (but usually still biased) samples are undertaken. At some point, despite the possible role of bias, decisions — such as Food and Drug Administration approval of a new drug — need to be made on the basis of biased samples. Little progress would be made in most fields if the results of all studies with biased samples were summarily dismissed.

At the same time, it is important to note that the statistical remedies for errors due to biased samples are extremely limited. Because we usually do not know the extent to which a particular bias has affected our results (e.g., we do not know how nonrespondents to a questionnaire would have answered the questions), it is usually not possible to adjust for errors created by bias. Thus, when biased samples are used, the results of statistical analyses of the data should be viewed with great caution.

Statistical results based on observations of biased samples should be viewed with great caution.

Exercise for Section 3

Factual Questions

1. What term is used to refer to all members of a group in which a scientist has an interest?

2. If samples yield statistics, what do populations yield?

3. What is the most important characteristic of a good sample?

4. If a scientist uses a sample of volunteers from a population, should we presume that the sample is biased?

5. What type of sampling eliminates bias?

6. Briefly describe how one could select a simple random sample.

7. Does random sampling produce sampling errors?

8. Is selecting a very large sample an effective way to reduce the effects of a serious bias in sampling?

Questions for Discussion

9. Are you convinced that using a rather small unbiased sample is better than using a very large biased sample? Why? Why not?

10. Be on the lookout for a news report of a scientific study in which a biased sample was used. If you find one, briefly describe it here.

Notes:

Section 4

Scales of Measurement

Scales of measurement (also known as *levels of measurement)* help investigators determine what type of statistical analysis is appropriate. As a consumer of research, you can often tell from the description of the measurement techniques what level was used and then judge whether the analysis was appropriate. It is important to master this section because it is referred to in a number of others that follow.

The lowest level of measurement is *nominal* (also known as *categorical)*. It is helpful to think of this level as the *naming* level. Here are some examples:

Subjects name the political parties with which they are affiliated.

Subjects name their gender.

Subjects name the states in which they reside.

Notice that the categories subjects name in these examples do not put the subjects in any particular order. There is no basis on which we could all agree for saying that Republicans are either higher or lower than Democrats. The same is true for gender and state of residence.

The next level of measurement is *ordinal*. Ordinal measurement puts subjects in rank *order* from high to low, but it does *not* indicate how much higher or lower one subject is in relation to another. To understand this level, consider these examples:

Subjects are ranked according to their height; the tallest subject is given a rank of 1, the next tallest is given a rank of 2, and so on.

The lowest level is *nominal*, which is the naming level.

Ordinal measurement puts subjects in rank order.

15

Three brands of hand lotion are ranked according to consumers' preferences for them.

In the examples above, the measurements tell us the relative standings of subjects but not the amount of difference among subjects. For example, we know that a subject with a rank of one is taller than a subject with a rank of two, but we do not know by how much. The first subject may be only one-quarter of an inch taller or may be two feet taller than the second.

The next two levels, *interval* and *ratio*, tell us by *how much* subjects differ. For example:

Interval and *ratio* scales measure how much subjects differ from each other.

The height of each subject is measured to the nearest inch.

The number of times each pigeon presses a button in the first minute after receiving a reward.

Notice that if one subject is 5'6" tall and another is 5'8" tall, we not only know the order of the subjects, but we also know by how much the subjects differ from each other. Both *interval* and *ratio* scales have equal intervals; for instance, the difference between one inch and two inches is the same as the difference between four inches and five inches.

In most statistical analyses, *interval* and *ratio* measurements are analyzed in the same way. There is a scientific difference, however. An *interval* scale does not have an absolute zero. For example, if we measure intelligence, we do not know exactly what constitutes zero intelligence and, thus, cannot measure it.[1] In contrast, a *ratio* scale has an absolute zero.[2] For example, we know where the zero point is on a tape measure when we measure height.

Ratio scales have an absolute zero; *interval* scales do not.

If you are having trouble mastering levels of measurement, first memorize this environmentally friendly phrase:

---

[1] Most applied researchers treat the scores from standardized tests (except percentile ranks and grade-equivalent scores) as *interval* scales of measurement, even though there is some controversy as to whether, for example, the difference between an IQ of 100 and 110 is the same as the difference between 110 and 120.

[2] Thus, the ratio scale is the only one in which we may take ratios of measurement. For example, we are permitted to make statements such as "John is twice as tall as Sam" (a ratio of 2 to 1) only when we use a ratio scale.

> **No Oil In Rivers**

The first letters of the words (NOIR) are the first letters in the names of the four levels of measurement in order from lowest to highest. Then read this section again and associate definitions with each type.

Exercise for Section 4

Factual Questions

1. What is the name of the lowest scale of measurement?

2. Which scale of measurement puts subjects in rank order?

3. Which two scales of measurement measure how much subjects differ from each other?

4. Which scale of measurement has an absolute zero?

5. If you measure the weight of subjects in pounds, which scale of measurement are you using?

6. If you rank employees from most cooperative (rank of 1) to least cooperative, which scale of measurement are you using?

7. If you ask subjects to name the country they were born in, which scale of measurement are you using?

8. What phrase should you memorize in order to remember the scales of measurement in order?

9. Which scale of measurement is between the ordinal and ratio scales?

Section 5

Descriptive and Inferential Statistics

Descriptive statistics summarize data.[1] For example, suppose you have the scores on a standardized test for 500 subjects. One way to summarize the data is to calculate an *average* score, which indicates how the typical person scored. You might also determine the highest and lowest scores, which would indicate how much the scores varied. These and other descriptive statistics are described in detail in this book.

Inferential statistics are tools that tell us how much confidence we can have when we generalize from a sample to a population.[2] You are familiar with national opinion polls in which a carefully drawn sample of only about 1,500 adults is used to estimate the opinions of the entire adult population of the United States. The pollster first calculates *descriptive statistics*, such as the *percentage* of respondents who are in favor of capital punishment and the percentage who are opposed. Having sampled, he or she knows that the results may not be accurate because the sample may not be representative; in fact, the pollster knows that there is a high probability that the results are off by at least a small amount. This is why pollsters often mention a *margin of error*, which is an inferential statistic. It is reported as a warning to the audience that random sampling may have produced errors, which should be considered when interpreting results.[3] For example, a weekly news magazine recently reported that in a national poll 58% of the respondents believed that the economy was improving; a footnote indicated that the margin of error was ±2.3. This means that the pollster was

> *Descriptive statistics* summarize data.
>
> An *average* is a descriptive statistic.
>
> *Inferential statistics* are tools that help us generalize from a sample to a population.
>
> A *percentage* is a descriptive statistic.
>
> A *margin of error* is an inferential statistic.

---

[1] Do not confuse *descriptive research* (see Section 2) with *descriptive statistics*. Descriptive research, like experimental research, employs both descriptive and inferential statistics.

[2] The word *inferential* comes from *infer*. When we generalize from a sample to a population, we are *inferring* that the sample is representative of the population.

[3] The measurement techniques, especially the wording of the question(s), may also produce errors. That is why sophisticated consumers of research usually want to know the exact wording of a question, especially if important decisions are to be made based on the results.

confident that the true percentage for the whole population was within 2.3 percentage points of 58%.[4]

You probably recall that a *population* is any group in which a scientist is interested. It may be large, such as all adults age 18 and over who reside in the United States, or it might be small, such as all registered nurses employed by a specific hospital. Scientists are free to choose populations of interest and should clearly define them when writing reports of their studies. A study in which all members of a population are included is called a *census*. A census is often feasible and desirable when working with small populations (e.g., an algebra teacher may wish to pretest all students at the beginning of a course, which will determine at what level to begin instruction and how to teach the class). When a population is large, it is more economical to use only a sample of the population. With modern sampling techniques, highly accurate information can be obtained using relatively small samples. Various methods of sampling are described later in this book. Note that inferential statistics are *not* needed when analyzing the results of a census because there is no sampling error.

> A *census* is a study in which all members of a population are included.

> *Inferential statistics* are not needed when analyzing the results of a census.

Exercise for Section 5

Factual Questions

1. Which type of statistics summarizes data?

2. Is an average a descriptive or inferential statistic?

3. Is a percentage a descriptive or inferential statistic?

4. Is a margin of error a descriptive or inferential statistic?

---

[4]Margins of error are described in detail in Section 16.

5. Which type of statistics helps us generalize from a sample to a population?

6. What is the name of the type of study in which all members of a population are included?

7. Why are inferential statistics not needed when analyzing the results of a census?

Notes:

Section 6

Frequencies, Percentages, and Proportions

A *frequency* is the number of subjects or cases; its symbol is f.[1] N, meaning *number of subjects*, is also used to stand for frequency.[2] Thus, if you see in a report that $f = 23$ for a score of 99, you know that 23 subjects had a score of 99. Likewise, if you see that $N = 23$ for a score of 99, you know that 23 subjects had a score of 99.

A *percentage*, whose symbol is P or %, indicates the number per hundred who have a certain characteristic. Thus, if you are told that 44% of the citizens in a town are registered as Democrats, you know that for each 100 registered voters, 44 are Democrats. To determine how many (the *frequency*) are Democrats, multiply the total number of registered voters by .44. Thus, if there are 2,313 registered voters, .44 x 2,313 = 1,017.72 are Democrats. In a report in the general media, this would probably be rounded to 1,018. In a report in an academic journal, thesis, or dissertation, however, the answer usually would be reported to two decimal places.

To calculate a percentage, use division. Consider this example: If 22 of 84 gifted children in a sample report being afraid of the dark, determine the percentage by dividing the number who are afraid by the total number of children and then multiply by 100; thus, $22 \div 84 = .2619$ x $100 = 26.19\%$. This result means that, based on the sample, if you questioned *100 subjects* from the same population, you would expect about 26 of them to report being afraid of the dark. Notice that only 84 subjects were actually studied, yet the result is still based on 100.

A *proportion* is part of one (1). In the previous paragraph, the proportion of children afraid of the dark is .2619 or .26 — the answer

---

A *frequency* is the number of cases; its symbol is f.

The symbol for *number of subjects* is N.

A *percentage* indicates the number per hundred.

To determine a percentage, divide the smaller number by the total and then multiply by 100.

A *proportion* is a part of one (1).

---

[1] Note that f is italicized. If you do not have the ability to type in italics, underline the symbol. This applies to almost all statistical symbols. Also, pay attention to the case (i.e., capitalization). A lowercase f stands for *frequency*; an uppercase F stands for another statistic described later in this book.

[2] An uppercase N should be used when describing a population; a lowercase n should be used when describing a sample. Scientists are not always careful to follow this convention.

obtained before multiplying by 100. This means that *twenty-six hundredths of each child* is afraid of the dark. As you can see, proportions are harder to interpret than percentages; thus, percentages are usually preferred to proportions in all types of reporting; however, do not be surprised if you occasionally encounter proportions in scientific writing.

When reporting percentages, it is a good idea to also report the underlying frequencies because percentages alone can sometimes be misleading or not provide sufficient information. For example, if you read that 8% of the foreign language students at a university were majoring in Russian, you would not have enough information to make informed decisions on how to staff the foreign language department and how many classes in Russian to offer. If you read that $f = 12$ (8%), based on a total of 150 foreign language students, you would know that 12 students need to be accommodated.

Percentages are especially helpful when comparing two or more groups of different sizes. Consider these data:

	College A	College B
Total number of foreign language students:	$N = 150$	$N = 350$
Russian majors:	$N = 12$ (8%)	$N = 15$ (4%)

Notice that the frequencies tell us that College B has more Russian majors; the percentages tell us that per 100 students, College A has more Russian majors. Thus, if College A and College B had the same number of foreign language students, we would expect to find twice as many Russian majors at College A than at College B because 8% is twice 4%.

> Report the underlying frequencies when reporting percentages.

Exercise for Section 6

Factual Questions

1. What does *frequency* mean?

2. What is the symbol for a frequency?

3. For what does *N* stand?

4. If 21% of kindergarten children are afraid of monsters, how many out of each 100 are afraid?

5. What statistic is a part of one (1)?

6. According to this section, are *percentages* or *proportions* easier to interpret?

7. Why should the underlying frequencies usually be reported with percentages?

Notes:

Section 7

Shapes of Distributions

We can see the shape of a distribution of a set of scores by examining a *frequency distribution*, which is a table that shows how many subjects have each score. Consider the frequency distribution in Table 7.1.

Table 7.1 Distribution of depression scores

X	f
22	1
21	3
20	4
19	8
18	5
17	2
16	0
15	1
$N =$	24

The frequency (i.e., f, which is the number of subjects) associated with each score (X) is shown. Examination indicates that most of the subjects are near the middle of the distribution (i.e., near a score of 19) and that the subjects are spread out on both sides of the middle with the frequencies tapering off.

The shape of a distribution is even clearer if we examine a *frequency polygon*, which is a drawing that shows how many subjects have each score. The same data shown in Table 7.1 above is shown in the frequency polygon in Figure 7.1 below. For example, the frequency distribution shows that 3 subjects had a score of 21; this same information is displayed in the frequency polygon. The high point in the polygon shows where most of the subjects are (in this case, at a score of 19). The tapering off around 19 illustrates how spread out the subjects are around the middle.

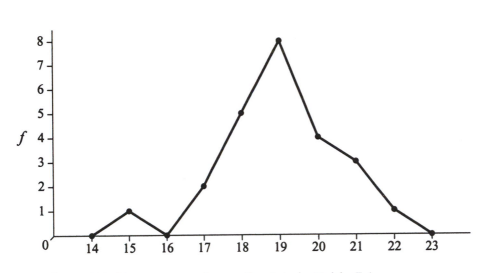

Figure 7.1 Frequency polygon for data in Table 7.1.

When there are many subjects, the shape of a polygon becomes smoother and is referred to as a *curve*. The most important shape is that of the *normal curve* — often called the bell-shaped curve — which is illustrated in Figure 7.2. It is important for two reasons. First, it is a shape very often found in nature. For example, the heights of women in a large population are normally distributed. There are small numbers of very short women, which is why the curve is low on the left; many women are of about average height, which is why the curve is high in the middle; and there are small numbers of very tall women. Here's another example: the average annual rainfall in Los Angeles over the past 105 years has been approximately normal. There have been a very small number of years in which there was extremely little rainfall, many years with about average rainfall, and a very small number of years with a great deal of rainfall. Another reason the normal curve is important is because it is used in inferential statistics, which are covered in the last part of this book.

The *normal curve* is the most important curve.

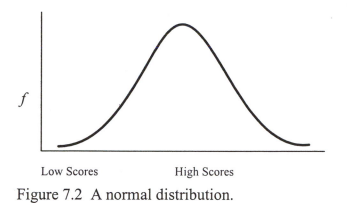

Low Scores   High Scores

Figure 7.2 A normal distribution.

Some distributions are *skewed*. For example, if you plot the distribution of income for a large population, you will find that it has a *positive skew* (i.e., is skewed to the right). Examine Figure 7.3. It indicates that there are large numbers of people with relatively small incomes; thus, the curve is high on the left. The curve drops off dramatically to the right forming a tail on the right; this tail is created by the small numbers of individuals with very high incomes. Skewed distributions are named for their tails. On a number line, positive numbers are to the right; hence, the term *positive skew*.

A distribution with a *positive skew* has a tail to the right.

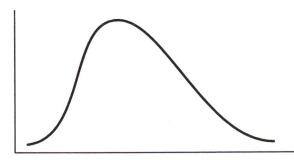

Figure 7.3   A distribution skewed to the right (positive skew).

When the tail is to the left, a distribution is said to have a *negative skew* (i.e., skewed to the left). See Figure 7.4. A negative skew would be formed if a large population of people was tested on skills in which they have been thoroughly trained. For example, if you tested a very large population of recent nursing school graduates on basic nursing skills, a distribution with a negative skew should emerge. There should be large numbers of graduates with high scores, but there should be a tail to the

A distribution with a *negative skew* has a tail to the left.

left created by a small number of nurses who, for one reason or another such as being physically ill on the day the test was administered, did not perform well on the test.

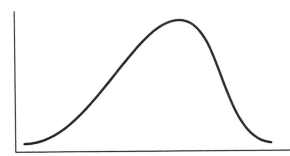

Figure 7.4 A distribution skewed to the left (negative skew).

Bimodal distributions have two high points. A curve such as that in Figure 7.5 is called bimodal even though the two high points are not equal in height. Such a curve is most likely to emerge when human intervention or a rare event has changed the composition of a population. For example, if a civil war in a country cost the lives of many young adults, the distribution of age after the war might be bimodal, with a dip in the middle. In general, they are less frequently found than other types of curves.

A *bimodal distribution* has two high points.

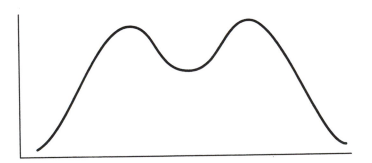

Figure 7.5 A bimodal distribution.

The shape of a distribution should always be examined before proceeding with additional statistical analyses because the shape has important implications for determining which average to compute — a topic that will be taken up next in this book. It is also important for determining which other statistics are appropriate to compute.

The shape of a distribution has implications for how data should be analyzed statistically.

Exercise for Section 7

Factual Questions

1. What is the name of a table that shows how many subjects had each score?

2. What does a frequency polygon show?

3. What is the most important type of curve?

4. In a distribution with a negative skew, is the tail to the left or to the right?

5. When plotted, income in large populations usually has what type of skew?

6. Suppose that on a 100-item multiple-choice test almost all students scored between 95 and 100 but a small scattering of subjects scored as low as 20. When plotted as a curve, the distribution will show what type of skew?

7. Suppose that a cross section of high school students took a very difficult scholarship examination and almost all did very poorly, but a small number managed to score very high. When plotted as a curve, the distribution will show what type of skew?

8. What is the name of the type of curve that has two high points?

Notes:

Section 8

The Mean: An Average

The most popular average is the *mean*.[1] It is so popular that is it sometimes simply called *the average*; however, this is an ambiguous term because several different types of averages are used in statistics.

Computation of the mean is quite simple; just sum (i.e., add up) the scores and divide by the number of scores. Here's an example:

Scores: 5, 6, 7, 10, 12, 15

Sum of scores: 55

Number of scores: 6

Computation of mean: $55/6 = 9.166 = 9.17$

Notice in the example that the answer was computed to three decimal places and rounded to two. In scientific work, the mean is usually reported to two decimal places.

There are several symbols for the mean. In scientific journals, the most commonly used symbols are M and m.[2] Throughout this book, they will be used as the symbols for the mean. Many statisticians, however, use this symbol:

$$\bar{X}$$

It is pronounced *X-bar*. Remember that X without the bar stands for a score or a set of scores.

An important characteristic of the mean is that it is the balance point in a distribution, that is, it is the *point around which all the deviations sum to zero*. The example in Table 8.1 illustrates this characteristic. The sum of the scores is 60; dividing this by 5 yields a mean of 12.00. By subtracting the mean from each score, we obtain the deviations from the

The most popular average is the mean.

To compute the mean, sum the scores and divide by the number of scores.

M or m are the most commonly used symbols for the mean in journals.

The mean is the balance point in a distribution.

---

[1] Its full, formal name is the *arithmetic mean*. Other averages are described in the next section.
[2] Strictly speaking, the uppercase M should be used when describing an entire population and the lowercase m should be used when describing a sample drawn from a population. Many authors of applied research, however, ignore this convention.

mean. These deviations sum to zero.[3] (Notice that the negatives cancel out the positives when summing.)

The deviations from the mean sum to zero.

Table 8.1 Scores and their deviations from their mean

Score	Mean	Deviation
7	12.00	-5
11	12.00	-1
11	12.00	-1
14	12.00	2
17	12.00	5
Sum of deviations = 0		

A major drawback of the mean is that it is drawn in the direction of extreme scores. This is a problem if there are *either* some extremely high scores that pull it up *or* some extremely low scores that pull it down. Here's an example of the contributions given to charity by two groups of children, expressed in cents:

The mean is pulled in the direction of extreme scores, which can be misleading.

Group A: 1, 1, 2, 3, 3, 4, 4, 4, 5, 5, 5, 5, 6, 6, 6, 7, 8, 10, 10, 10, 11
Mean for Group A = 5.52

Group B: 1, 2, 2, 3, 3, 3, 4, 4, 5, 5, 5, 6, 6, 6, 6, 6, 9, 10, 10, 150, 200
Mean for Group B = 21.24

Notice that overall, the two distributions are quite similar. Yet the mean for Group B is much higher than the mean for Group A because of two students who gave extremely high contributions of 150 cents and 200 cents.[4] If only the means for the two groups were reported without reporting all of the individual contributions, it would suggest that the average student in Group B gave about 21 cents when, in fact, none of the

---

[3]This is a *defining characteristic* of the mean. There is only one value that has this characteristic for a given distribution. Any value that does *not* have this characteristic is *not* the mean. Note that if the mean is not a whole number, the sum of the deviations may vary slightly from zero because of rounding when determining the mean because a rounded mean is not *precisely* accurate.
[4]A distribution with extremes that produce a tail in one direction but not the other is called *skewed*. See Section 7.

students made a contribution of about this amount. An average that provides a more accurate indication of the center for this situation is described in the next section.

Another limitation of the mean is that it is appropriate for use only with *interval* and *ratio* scales of measurement (see Section 4) — unlike the averages described in the next section.

Note that a synonym for *average* is *measure of central tendency*. Although the latter term is seldom used in reports of scientific research, you may encounter it if you refer to other statistics texts.

The mean should be used only with *interval* and *ratio* scales of measurement.

A synonym for *average* is *measure of central tendency*.

Exercise for Section 8

Factual Questions

1. How is the mean computed?

2. What are the most commonly used symbols for the mean (in scientific journals)?

3. For a given distribution, if you subtract the mean from each score to get deviations and then sum the deviations, what will the sum of the deviations equal?

4. If most subjects have similar scores but there are a few very high scores, what effect will the very high scores have on the mean?

5. With what scales of measurement is the mean associated?

6. The term *measures of central tendency* is synonymous with what other term?

Notes:

Section 9

Mean, Median, and Mode

The *mean*, described in the previous section, is the *balance point* in a distribution. It is the most frequently used average.[1]

The *mean* is the balance point in a distribution.

An alternative average is the *median*. It is the value in a distribution that has 50% of the cases above it and 50% of the cases below it. Thus, it is the *middle point* in a distribution. In Example 1, there are 11 scores. The middle score, with 50% on each side, is 81, which is the median. Note that there are five scores above 81 and five scores below 81.[2]

The *median* is the *middle point* in a distribution.

Example 1: Scores (arranged in order from low to high):
61, 61, 72, 77, 80, 81, 82, 85, 89, 90, 92

In Example 2, there are 6 scores. Because there is an even number of scores, the median is halfway between these scores. To find this value, sum the two middle scores (7 + 10 = 17) and divide by 2 (17/2 = 8.5). Thus, 8.5 is the median of this set of scores.

Example 2: Scores:
3, 3, 7, 10, 12, 15

An advantage of the median is that it is insensitive to extreme scores.[3] This is illustrated by Example 3, in which the extremely high score of 229 has no effect on the value of the median. The median is 8.5, the same value as in Example 2, despite the extremely high score.

The median is insensitive to extreme scores.

Example 3: Scores:
3, 3, 7, 10, 12, 229

---

[1] Another term for *average* is *measure of central tendency*.

[2] When there are tie scores in the middle, that is, when the middle score is earned by more than one subject, this method for determining the median is only approximate and the result should be referred to as an *approximate median*.

[3] In Section 8, it was noted that the mean is pulled in the direction of extreme scores, which may make it a misleading average for skewed distributions.

The *mode* is another average. It is simply the *most frequently occurring score*. In Example 4, the mode is 7 because it occurs more often than any other score.

Example 4: Scores:
2, 2, 4, 6, 7, 7, 7, 9, 10, 12

A disadvantage of the mode is that there may be more than one mode for a given distribution. This is the case in Example 5 in which both 20 and 23 are modes.

Example 5: Scores:
17, 19, 20, 20, 22, 23, 23, 28

Here are some guidelines to use when choosing an average.

(1) Other things being equal, choose the mean because more powerful statistical tests described later in this book can be applied to it than to the other averages. However, (a) the mean is appropriate only for describing approximately symmetrical distributions; it is inappropriate for describing highly skewed distributions, and (b) the mean is only appropriate for describing interval and ratio data. (See Section 4.)

(2) Choose the median when the mean is inappropriate. The exception to this guideline is when describing nominal data. Nominal data (see Section 4) are naming data such as political affiliation, ethnicity, etc. There is no natural order to these data; therefore, they cannot be put in order, which is required in order to calculate the median.

(3) Choose the mode when an average is needed to describe nominal data. Note that when describing nominal data, it is often not necessary to use an average. For example, if there are more Democrats than Republicans in a community, the best way to describe this is to report the percentage of people registered in each party. To simply state that the modal political affiliation is Democratic (which is the

----

The *mode* is the *most frequently occurring score.*

More powerful statistical tests can be applied to the mean than to other averages.

The mean is appropriate only for use with approximately symmetrical distributions and for interval and ratio data.

Choose the median when the mean is inappropriate, except when analyzing nominal data.

Choose the mode as the average for nominal data.

For nominal data, an average may not be needed.

average in this case) is much less informative than reporting percentages.

Note that in a perfectly symmetrical distribution such as the normal distribution, the mean, median, and mode all have the same value. In skewed distributions, their values are different as illustrated in the following figures. In a distribution with a positive skew, the mean has the highest value because it is pulled in the direction of the extremely high scores. In a distribution with a negative skew, the mean has the lowest value because it is pulled in the direction of the extremely low scores. As noted earlier, do not use the mean when a distribution is highly skewed.

In the normal distribution, the mean, median, and mode have the same value.

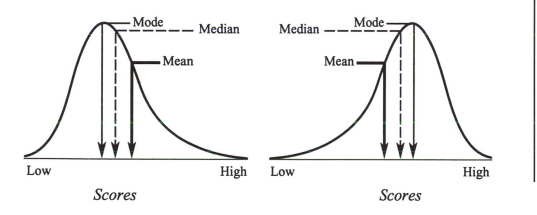

When there is a positive skew, the *mean* is higher than the *median*.

When there is a negative skew, the *median* is higher than the *mean*.

Exercise for Section 9

Factual Questions

1. Which average always has 50% of the cases below it?

2. Which average is defined as the most frequently occurring score?

3. If you read that the median for a group of subjects equals 42 on a test, what percentage of the subjects had scores higher than 42?

4. What is the mode of the following scores?   11, 13, 15, 15, 17, 21, 25

5. If the mean is inappropriate for describing a set of scores measured at the interval level, which average should be used?

6. In a distribution with a positive skew, does the mean or median have a higher value?

7. In a distribution with a negative skew, does the mean or median have a higher value?

Section 10

Range and Interquartile Range

Variability refers to the differences among scores, which indicate how subjects vary. In scientific writing, most authors report a statistic designed to indicate the amount of variability[1] immediately after reporting an average. Synonyms for *variability* are *spread* and *dispersion*.

How much subjects in a group vary is important for statistical and practical reasons. Suppose, for example, you were going to teach fourth grade next year and were offered a choice between two classes—both of which were very similar in terms of their average scores obtained on a standardized test. Before making a choice, you would be wise to ask about the variability. Suppose you learned that one class had very little variability — their scores were all very close to their average — and that the other had tremendous variability — their scores varied from the highest to the lowest possible with a great deal of spread in between. Which class would you choose? There is no right or wrong answer to the question, but clearly information on variability would be important in helping you make a decision.

A simple statistic that describes variability is the *range*. It is the difference between the highest score and the lowest score.[2] For the scores in Example 1, the range is 18 (20 minus 2). One could report the 18 as the range or simply state that the scores range from 2 to 20.

Example 1: Scores:
2, 5, 7, 7, 8, 8, 10, 12, 12, 15, 17, 20

A weakness of the range is that it is based on only the two most extreme scores, which may not reflect the true variability in the entire group. Consider Example 2. As in Example 1, the range is also 18. However, there is much less variability among the subjects than in Example

Variability refers to differences among scores.

Synonyms for *variability* are *spread* and *dispersion*.

The *range* is the difference between the highest score and the lowest score.

A weakness of the range is that it is based on the two most extreme scores.

---

[1] This group of statistics is often called *measures of variability.*
[2] Some statisticians add the constant one (1) to the difference when computing the range.

1; in Example 2 one subject with an extremely high score relative to the group (i.e., 20) has had an undue influence on the range.

Example 2: Scores:
2, 2, 2, 3, 4, 4, 5, 5, 5, 6, 6, 20

Scores such as the 20 in Example 2 are known as *outliers*. They lie far outside the range of the vast majority of other scores.

A better measure of variability is the *interquartile range* (*IQR*). It is defined as the range of the middle 50% of the subjects. By using only the middle 50%, we are describing the range of the majority of the subjects and, at the same time, are ignoring outliers that may have an undue influence on the ordinary *range*. Example 3 illustrates the meaning of the *interquartile range*. Notice that the scores are in order from low to high. The arrow on the left separates the lowest 25% from the middle 50%, and the arrow on the right separates the highest 25% from the middle 50%. It turns out that the range for the middle 50% is 3 points.[3] When you report 3.0 to an audience, they will know that the range of the middle 50% of subjects is only 3 points. Note that the undue influence of the outlier of 20 has been overcome by using the *interquartile range*.

Example 3: Scores:
2, 2, 2, 3, 4, 4, 5, 5, 5, 6, 7, 20
⇑   ⇑

The interquartile range may be thought of as a first cousin of the *median*.[4] Thus, when the *median* is reported as the average for a set of scores, it is customary to report the *interquartile range* as the measure of variability.[5]

Outliers are scores that lie far outside the range of the vast majority of scores.

The *interquartile range* (*IQR*) is the range of the middle 50% of the subjects.

The *IQR* is better than the *range* because it ignores outliers.

When the *median* is reported as the average, the *IQR* is usually reported for variability.

---

[3]For those of you interested in the computation of the *IQR*, notice that the right arrow is at 5.5 and the left arrow is at 2.5. By subtracting (5.5 - 2.5 = 3.0), you obtain the *IQR*.

[4]To calculate the median, we count to the middle of the distribution. To calculate the *IQR*, we count off the top and bottom quarters. This similarity in computations illustrates why they are cousins.

[5]The measure of variability that is associated with the mean is introduced in the next section. See the previous section for guidelines on when to report the median and the mean.

Exercise for Section 10

Factual Questions

1. What are the two synonyms for *variability*?

2. If the differences among a set of scores is great, do we say that there is *much variability* or *little variability*?

3. What is the definition of the range?

4. What is a weakness of the range?

5. What is the outlier in the following set of scores?   2, 15, 16, 16, 17 18, 20

6. What is the definition of the interquartile range?

7. Is the interquartile range affected by outliers?

8. If the interquartile range is 10 points for a set of scores, what percentage of the subjects have scores within 10 points of each other?

9. When the median is reported as the average, which measure of variability should be reported?

Notes:

Section 11

Standard Deviation

The *standard deviation* is the most popular measure of *variability*. As you learned in the previous section, *variability* refers to the differences among scores, which indicate how subjects vary. In scientific writing, most authors report a statistic designed to indicate the amount of variability[1] immediately after reporting an average. Synonyms for *variability* are *spread* and *dispersion*.

The standard deviation measures how much subjects differ from the *mean* of their group. It is a special type of average of the deviations of the scores from their mean.[2] Thus, the more spread out subjects are around their mean, the larger the standard deviation. Comparison of Examples 1 and 2 illustrate this principle, where S is the symbol for the standard deviation.[3] Notice that the mean is the same for both groups, but the group with the greater variability (i.e., Group A) has a larger standard deviation.

Example 1:
Scores for Group A: 0, 0, 5, 5, 10, 15, 15, 20, 20
$M = 10.00, S = 7.45$

Example 2:
Scores for Group B: 8, 8, 9, 9, 10, 11, 11, 12, 12
$M = 10.00, S = 1.49$

Now consider the scores of Group C in Example 3. All subjects have the same score; therefore, there is no variability. When this is the case, the standard deviation equals zero, which indicates the complete lack of variability.

The *standard deviation* is the most popular measure of *variability*.

Synonyms for *variability* are *spread* and *dispersion*.

The standard deviation is a special type of average of the deviations of the scores from their *mean*.

S is the symbol for the standard deviation.

Group A has a larger standard deviation than Group B. Thus, Group A's scores are more variable.

---

[1] This group of statistics is often called *measures of variability*.
[2] For those of you who are mathematically inclined, Appendix A shows how to compute the standard deviation. By studying this appendix, you will learn what is meant by a "special type of average of the deviations."
[3] The lowercase *s* is used when the subjects are only a sample of a population. Applied researchers often use *S.D.* and *s.d.* as symbols for the standard deviation.

Example 3:

Scores for Group C: 10, 10, 10, 10, 10, 10, 10, 10, 10, 10

$M = 10.00, S = 0.00$

There is no variability in Group C's scores; thus, the standard deviation equals zero.

Considering the three examples, it is clear that the more subjects differ from their mean, the larger the standard deviations. Conversely, the less they differ from their mean, the smaller the standard deviations.

Let's review; even though the three groups have the same mean:

(1) Group C has no variability.

(2) Group A has more variability than Groups B and C.

(3) Group B has more variability than Group C.

Thus, if you were reading a research report about the three groups, you would obtain important information about them by considering their standard deviations.

The standard deviation takes on a special meaning when considered in relation to the normal curve (see Section 7) because it was designed expressly to describe this distribution. Here's a simple rule to remember: *about two-thirds of the cases lie within one standard deviation unit of the mean in a normal distribution.* (Note that "within one standard deviation unit" means one unit on *both* sides of the mean.) For example, suppose that the mean of a set of normally distributed scores equals 70.00 and the standard deviation equals 10.00. Then, about two-thirds of the cases lie within 10 points of the mean. More precisely, 68% (a little more than two-thirds) of the cases lie within 10 points of the mean, which is illustrated in Figure 11.1.

About two-thirds of the cases lie within one standard deviation unit of the mean.

More precisely, 68% (a bit more than 2/3) lies within one standard deviation unit of the mean.

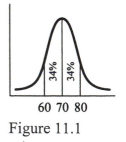

60 70 80

Figure 11.1

Suppose for another group, the mean of their normal distribution equals 25.00 and their standard deviation equals 5.00. Then, 68% of the cases lies within 5 points of the mean, as illustrated in Figure 11.2.

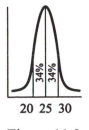

20 25 30

Figure 11.2

At first, this seems like magic — regardless of the value of the mean and standard deviation, 68% of the cases lie within one standard deviation unit in a normal curve. Actually, it is not magic but a property of the normal curve. When a scientist is calculating the standard deviation for a normal distribution, she is actually calculating the number of points she needs to go out from the mean in both directions to locate the middle 68% of the cases.[4] This two-thirds rule does *not* strictly apply if the distribution is *not* normal. The less normal it is, the less accurate the rule is.

The two-thirds rule is a property of the normal curve.

By now, you have probably figured out the standard deviation is a first cousin of the mean. Thus, when scientists report the mean (the most popular average), they usually also report the standard deviation.

The two-thirds rule does not strictly apply if the distribution is not normal.

When the mean is reported, the standard deviation is also usually reported.

Exercise for Section 11

Factual Questions

1. The standard deviation is based on how much subjects differ from what other statistic?

2. If the differences among a set of scores is small, do we say that there is *much variability* or *little variability*?

---

[4]When you go out one standard deviation on both sides of the mean, you reach the *points of inflection* — the points where the curve changes direction and begins to go out more quickly than it goes down.

3. What is the symbol for the standard deviation?

4. Will the scores for Group D or Group E below have a larger standard deviation if it were computed? (Do *not* compute the standard deviations; examine the scores.)

    Group D: 20, 21, 22, 24, 24, 27, 27   Group E: 15, 19, 20, 21, 25, 28, 32

5. If everyone in a group has the same score, what will be the value of the standard deviation of the scores?

6. If you found the following statistics in a research report, which group should you conclude has the greatest variability?

    Group F: $M = 30.23$, $S = 2.14$
    Group G: $M = 25.99$, $S = 3.01$
    Group H: $M = 22.43$, $S = 1.79$

7. What percentage of the cases in a normal curve lies within one standard deviation unit of the mean (i.e., between one standard deviation unit above the mean and one standard deviation unit below the mean)?

8. Suppose $M = 30.00$ and $S = 2.00$ for a normal distribution of scores. What percentage of the cases lies between scores of 28 and 32?

9. Suppose $M = 100.00$ and $S = 15.00$ for a normal distribution of scores. About 68% of the cases lie between what two scores?

10. The standard deviation is usually reported in conjunction with what other statistic?

Section 12

Correlation

Correlation refers to the extent to which two variables are related across a group of subjects. Consider scores on the College Entrance Examination Board's *Scholastic Aptitude Test* (*SAT*) and first-year GPA in college. Because the *SAT* is widely used as a predictor in college student selection, there should be a correlation between these scores and GPA. Consider Example 1 in which *SAT-V* refers to the verbal portion of the *SAT*.[1] Is there a relationship?

Example 1:

Student	SAT-V	GPA
John	333	1.0
Janet	756	3.8
Thomas	444	1.9
Scotty	629	3.2
Diana	501	2.3
Hillary	245	0.4

Indeed, there is. Notice that students who scored high on the *SAT-V* such as Janet and Scotty had the highest GPAs. Also, those who scored low on the *SAT-V* such as Hillary and John had the lowest GPAs. This type of relationship is called *direct* or *positive*. In a direct relationship, those who score high on one variable tend to score high on the other, *and* those who score low on one variable tend to score low on the other.

Example 2 shows the relationship between a personality scale that measures willingness to take orders (on a scale from 0 to 20, where 20 represents eagerness to take orders) and the number of days served in county jail for misdemeanors. Is there a relationship? If you consider the scores carefully, you will see that there is. Notice that those who have a

---

[1] *SAT-V* scores range from 200 to 800.

high willingness to take orders such as Jake and David served the fewest days in jail, while inmates such as Jason and Joan who have low willingness served the most days in jail. Such a relationship is called *inverse* or *negative*. In an inverse relationship, those who score high on one variable tend to score low on the other.

Example 2:

Inmate	Willingness	Days Served
Jake	20	3
Jason	1	65
Sarah	10	20
Dick	11	24
David	18	7
Joan	3	40

It is important to note that just because we have established a correlation, we have not necessarily established a *causal relationship*. In a causal relationship, one variable is found to cause a change in the other — that is, one variable is found to affect the other. Consider the hypothetical relationship between willingness to take orders and days served in jail. Although a relationship was found, there could be many *causal* explanations. For example, those who have better attorneys may have had better instruction on how to be compliant (and, thus willing to take orders) and those same prisoners' attorneys, because they are better, may have had greater success in having them released from jail early. Many other explanations may be possible.

In order to determine *cause-and-effect*, a controlled *experiment* is needed in which different treatments are tried. (See Section 2.) For example, if a treatment given to an experimental group is shown to lead to a change not found in a comparable control group, then we would have evidence regarding causality. Note that in Examples 1 and 2 above, two variables were measured but no treatments were given. Even though we generally should not infer causality from a correlational study, we are still often interested in correlation. For example, the College Board and

In an *inverse* or *negative* relationship, those who score high on one variable tend to score low on the other.

Establishing a correlation does not necessarily establish a *causal relationship*.

In order to determine *cause-and-effect*, a controlled *experiment* is needed.

the colleges that use its test are interested in how well the test works in predicting success in college; it is not necessary to show what causes high GPAs in college in order to make the test useful. Also, correlations are of interest in developing theories. Often a postulate of a theory may say that *X* should be related to *Y*. If a correlation is found, it helps to support the theory.

Up to this point, we have examined the scores of only small numbers of subjects in clear-cut cases. However, in practice, large numbers of subjects are usually examined, which almost always results in exceptions to the trend. Consider Example 3, in which we have the same students as in Example 1 but with the addition of two others — Joe and Patricia.

Example 3:

Student	SAT-V	GPA
John	333	1.0
Janet	756	3.8
Thomas	444	1.9
Scotty	629	3.2
Diana	501	2.3
Hillary	245	0.4
Joe	630	0.9
Patricia	404	3.1

Joe has a high *SAT-V* score but a very low GPA; he is an *exception* to the rule that high values on one variable are associated with high values on the other. There may be a variety of explanations for this exception —Joe may have had a family crisis during his first year in college *or* he may have abandoned his good work habits to make time for TV viewing and campus parties as soon as he moved away from home to college. Patricia is another exception — perhaps she made an extra effort to apply herself to college work, which could not be predicted by the *SAT*. When studying hundreds of subjects, there will be many exceptions — some large and some small. To make sense of such data, statistical techniques are required. These will be explored in the next two sections.

When a large number of cases are examined, there are usually exceptions to the trend.

To make sense of data with many exceptions, statistical techniques are required.

Exercise for Section 12

Factual Questions

1. A direct relationship was found between scores on an algebra readiness test given before students took algebra and their first semester grades in algebra. This means that those who scored low on the readiness test tended to have what kind of grade?

2. What is a synonym for *direct*?

3. Is the relationship between the scores on Test X and Test Y direct or inverse?

Subject	Test X	Test Y
David	30	5
Julie	20	6
Happy	40	4
Shorty	50	3
Marcia	70	1
Kelly	60	2

4. If there is an inverse relationship, those who tend to score high on one variable tend to have what kind of score on the other variable?

5. If there is a direct relationship, those who tend to score high on one variable tend to have what kind of score on the other variable?

6. What type of study is needed in order to identify *cause-and-effect* relationships?

7. Is *correlation* a good way to determine *cause-and-effect*?

8. When a large number of cases are examined and a positive relationship is found, what else should one expect to find?

Questions for Discussion

9. Name two variables that you think have a direct relationship with each other.

10. Name two variables that you think have an inverse relationship with each other.

Notes:

Section 13

Pearson *r*

A statistician named Pearson developed a widely used statistic for describing the relationship between two variables. His statistic is often simply called the *Pearson r*. Its full, formal name is the *Pearson product-moment correlation coefficient,* and you might find variations on this in the literature such as *Pearson correlation coefficient* or the *product-moment correlation coefficient.* We'll begin by considering some basic properties of a *Pearson r*:

The full name of the *Pearson r* is *Pearson product-moment correlation coefficient.*

(1) It can range only from -1.00 to 1.00.
(2) -1.00 indicates a perfect inverse relationship—the strongest possible inverse relationship.
(3) 1.00 indicates a perfect positive relationship—the strongest possible direct relationship.
(4) 0.00 indicates the complete absence of a relationship.
(5) The closer a value is to 0.00, the weaker the relationship.
(6) The closer a value is to -1.00 or 1.00, the stronger it is.

A *Pearson r* ranges from -1.00 to 1.00.

Both -1.00 and 1.00 indicate a perfect relationship.

A value of 0.00 indicates the complete absence of a relationship.

Thus,

-1.00				0.00				1.00
⇑	⇑	⇑	⇑	⇑	⇑	⇑	⇑	⇑
perfect	strong	moderate	weak	none	weak	moderate	strong	perfect

Notice the labels *strong*, *moderate*, and *weak* are used in conjunction with both positive and negative values of *r*. Also, exact numerical values are not given for these labels. This is because the interpretation and labeling of an *r* may vary from one investigator to another and from one type of investigation to another. For example, one way to examine *test reliability* is to administer the same test twice to a group of subjects

The interpretation of a Pearson *r* varies, depending on the type of study.

55

without trying to change the subjects between administrations of the test. This will result in two scores per person, which can be correlated using Pearson's *r*. In such a study, a professionally constructed test should yield high values of *r* such as .85 or higher.[1] A result such as .65 probably would be characterized as only moderately strong. In another type of study, where high values of *r* are seldom achieved (such as predicting college GPAs from College Board scores earned a year earlier), a .65 might be interpreted as strong or even very strong.[2]

The interpretation of the values of *r* is further complicated by the fact that an *r* is *not a proportion*. Thus, an *r* of .50 is not half of anything. It follows that multiplying .50 by 100 does *not* yield a percentage; that is, .50 is *not* equivalent to 50%. This is important because we are used to thinking of .50 as being halfway between 0.00 and 1.00. In Section 14, you will learn how to compute another statistic that is directly related to *r* but that may be interpreted as a proportion and converted to a percentage.

An *r* is *not* a proportion. Thus, multiplying it by 100 does *not* yield a percentage.

Consult Appendix B for some additional notes on the interpretation of *r*.

Exercise for Section 13

Factual Questions

1. What is the full name of the Pearson *r*?

2. What does a Pearson *r* of 0.00 indicate?

3. What does a Pearson *r* of -1.00 indicate?

---

[1] A test is said to be reliable if its results are consistent. For example, if you measured the length of a table twice with a tape measure, you would expect similar results both times—unless your measurement technique was unreliable.

[2] Careful study of the literature on the topic being investigated is needed in order to arrive at a nonnumerical label or interpretation of a *Pearson r* that will be accepted by one's colleagues.

4. Which of the following indicates the strongest relationship?
   A. 0.00 B. 1.00 C. -.98 D. .50

5. Which of the following indicates the weakest relationship?
   A. .33 B. -.88 C. -1.00 D. .97

6. Is it possible for a relationship to be both direct and weak?

7. Is it possible for a relationship to be both inverse and strong?

8. Is a Pearson *r* a proportion that can be converted to a percentage by multiplying by 100?

Questions for Discussion

9. Name two variables that you believe would yield a positive value of *r* if their values were correlated. State whether you think the value of *r* would be high or low. (*Sample answer*: Weight and height should be positive, and the value of *r* should be fairly high.)

10. Very briefly describe a study you would like to conduct in which it would be appropriate to compute a Pearson *r*. Predict whether the *r* would be positive or negative and whether it will be high or low in value.

Notes:

Section 14

Coefficient of Determination

The *coefficient of determination* is useful when interpreting a Pearson *r*. Its symbol, r^2, explains how it is computed; to obtain it, simply square *r*. Thus, for a Pearson *r* of .60, r^2 equals .36 (.60 x .60 = .36).

Although the computation is simple, what it indicates is sometimes difficult for students to grasp. Let's begin by considering the scores in Table 14.1. Five young children took an oral vocabulary knowledge test

Table 14.1 Scores on two tests

Student	Vocabulary	Reading
John	3	6
Janet	5	8
Thomas	4	9
Scotty	9	10
Diana	10	12

before they began learning how to read. After six months of instruction, they took a reading test. As you can see, there is a positive relationship because those who are low on vocabulary (such as John) are also low on reading *and* those who are high on vocabulary (such as Diana) are also high on reading. Thus, we can say that the vocabulary scores are predictive of the subsequent reading scores. But how predictive? As it turns out, the value of the Pearson *r* for these scores is .90.[1] Thus, we can say that it is highly predictive. However, we can be more precise if we use the coefficient of determination. Let's consider how.

First, notice that there are differences among the scores on the vocabulary test; this is referred to as *variance*. There is also variance in the scores on the reading test. When interpreting a Pearson *r*, an important

> To obtain the *coefficient of determination,* square *r*.

> The *coefficient of determination* allows us to be precise when interpreting correlation coefficients.

---

[1]Computational procedures for obtaining the value of a Pearson *r* are beyond the scope of this book. In this example, *r* does not equal 1.00 because there are exceptions to the positive trend; notice that although Janet is higher than Thomas on vocabulary, she is lower than Thomas on reading.

question is: *what percentage of the variance on one variable is accounted for by the variance on the other?* If we are trying to predict reading scores from vocabulary scores, the question might be phrased as: *what percentage of the variance in reading is <u>predicted</u> by the variance on vocabulary?* The answer to the question is determined simply by computing r^2 and multiplying it by 100. For the scores shown in Table 14.1, $r = .90$. Thus,

$$.90 \times .90 = .81 \times 100 = 81\%$$

We have determined that 81% (*not* 90%) of the variance on one variable is accounted for by the variance on the other in this example.[2]

Let's put the 81% in perspective. Suppose we are trying to predict how subjects will score in reading. Suppose we naively put all of the subjects' names on slips of paper in a hat and draw a name and declare that the first name drawn will probably perform best on the reading test, and draw a second name and declare that this person will probably perform second best on the reading test, and so on. What percentage of the variance in reading will we predict using this procedure? In the long run with large numbers of subjects, the answer is about zero (0.00) percent. In the above example, the differences (i.e., variance) in vocabulary accounted for 81% of the differences (i.e., variance) in reading, which is greatly better than using a random process to make predictions.

It follows, however, that if we can account for 81% of the variance, 19% (100% − 81% = 19%) of the variance is *not* accounted for. Thus, there is much room for improvement in our ability to predict.

Consider Table 14.2. It shows selected values of r, the corresponding values of r^2, and the percentage of variance accounted for and not accounted for. Notice that small values of r shrink dramatically when converted to r^2, indicating that we should be very cautious when interpreting small values of r — they are farther from perfection than they might seem at first.[3]

If you are having difficulty understanding the coefficient of determination, consider again what Table 14.2 tells us. When $r = .10$, our ability to predict is 1% better than no ability; when $r = .20$, our ability to

---

The *coefficient of determination*, when converted to a percentage, tells us how much variance on one variable is accounted for by the variance on the other.

A percentage may be obtained by multiplying r^2 by 100.

If you draw names at random, your ability to predict is zero percent.

If $r = .90$, the ability to predict is 81% better than zero.

Table 14.2 shows the values of the *coefficient of determination* that correspond to selected values of r.

Small values of r shrink dramatically when squared.

---

[2]*Variance accounted for* is sometimes called *explained variance*.
[3]Note that if there is no variance on either variable, the Pearson r will equal 0.00 and r^2 will also equal 0.00. This is easy to see by example. For instance, suppose a group of students all had identical vocabulary scores. Since these scores fail to differentiate among students, they cannot predict who will have low reading scores, who will have average reading scores, etc.

Table 14.2   Selected values of r and r^2.

r	r^2	% accounted for	% *not* accounted for
.10	.01	1%	99%
.20	.04	4%	96%
.30	.09	9%	91%
.40	.16	16%	84%
.50	.25	25%	75%
.60	.36	36%	64%
.70	.49	49%	51%
.80	.64	64%	36%
.90	.81	81%	19%
1.00	1.00	100%	0%

predict is 4% better than no ability, etc. Thus, the coefficient of determination, when converted to a percentage, tells us how effective one variable is in predicting another.

Keeping Table 14.2 in mind when reading scholarly articles should give you pause because many authors interpret r's in the .20 to .40 range as indicating important relationships. In fact, they may be of some practical importance under certain circumstances. However, keep in mind that in this range, 84% to 96% of the variance on one variable is *not* accounted for by the other. When considering a prediction study, an r of .40 leaves much room for improvement when attempting to predict one variable from another.

When the values of r are less than .40, more than 84% of the variance is *not* accounted for.

Exercise for Section 14

Factual Questions

1. If you have a value of r, how do you determine the value of the coefficient of determination?

2. What is the symbol for the coefficient of determination?

3. When $r = .50$, what is the value of the coefficient of determination?

4. When $r = .50$, what percentage of the variance on one variable is accounted for by the variance on the other?

5. When $r = .50$, what percentage of the variance on one variable is *not* accounted for by the variance on the other?

6. Do large values or small values of r shrink more dramatically when squared?

7. When $r = .10$, is the percentage accounted for equal to 10%? Explain.

Section 15

Sampling Concepts

As noted in Section 3, an unbiased sample may be obtained by using *simple random sampling*. Putting names on slips of paper and drawing the number needed for the sample is a classic method of obtaining such a sample. For larger populations, it is more efficient to use a *table of random numbers*, a portion of which is shown in Appendix C.[1] In this table, there is no sequence to the numbers and, in a large table, each number appears about the same number of times. To use the table, first assign everyone in the population a *number name*. For example, if there are 90 people in the population, name the first person 01, the second person 02, the third person 03, etc., until you reach the last person whose number is 90.[2] (Often, computerized records have the individuals already numbered, which simplifies the process. Any set of numbers will work as number names as long as each one has a different number and each one has the same number of digits in his or her name.) To use the table, flip to any page in a book of random numbers and put your finger on the page without looking; this will determine your starting point. We'll start in the upper-left hand corner of the table in Appendix C for the sake of illustration. Because each person has a two-digit number name, the first two digits identify our first subject; this is person number 21. The next two digits to the right (ignoring the spaces between the columns which are provided only as a visual aid while reading the table) are 0 and 4; thus, person number 04 will also be included in our sample. The third number is 98. Because there are only 90 in the population, skip 98 and continue to the right to 08, which is the number of the next person drawn. Continue moving across the rows to select the sample.

Stratified random sampling is usually superior to *simple random sampling*. In this technique, the population is first divided into strata that

> Putting names on slips of paper and drawing them is a way to obtain a *simple random sample*.
>
> A *table of random numbers* may also be used to draw a random sample.
>
> To use a *table of random numbers*, give each person in the population a *number name*.

---

[1] Academic libraries have books of random numbers. Statistical computer programs can also generate them.
[2] The number of digits in the number names must equal the number of digits in the population total. For example, if there are 500 people in the population, there are 3 digits in the total, and there must be 3 digits in each name. Thus, the first case in the population is named 001.

are believed to be relevant to the variable(s) being studied. Suppose, for example, you wished to conduct a survey of the opinions on "date rape" held by all students on a college campus. If you suspect that males and females might differ in their opinions, it would be desirable to first stratify the population according to gender and then draw separately from each stratum at random. Specifically, you would draw a random sample of males and separately draw a random sample of females. The same percentage should be drawn from each stratum. For example, if you want to sample 10% of the population and there are 1,600 males and 2,000 females, you would draw 160 males and 200 females. Notice that there are more females in the sample than males, which is appropriate because the females are more numerous in the population. It is important to notice that we are *not* stratifying in order to compare males with females; rather we are attempting to obtain a sample of the entire college population that is representative in terms of gender.[3] With stratified random sampling, we have retained the benefits of randomization (i.e., the elimination of bias) and have gained the advantage of having appropriate proportions of males and females.[4] If your hunch was correct that males and females differ, you would have increased the precision of your results by stratifying.

> In *stratified random sampling*, draw subjects at random separately from each stratum.

> Draw the same percentage, not the same number, from each stratum.

Note that stratifying does not eliminate sampling errors. For example, when you drew the females at random, you may have, by chance, obtained females for your sample that are not representative of all females on the campus; the same, of course, holds true for men. However, you have eliminated sampling errors associated with gender because if you had used *simple random sampling* you could have obtained a disproportionately large number of either males or females.

For large-scale studies, *multistage random sampling* may be used. In this technique, you might draw a sample of counties at random from all counties in the country, then draw voting precincts at random from all precincts in the counties selected, and finally draw individual voters at random from all precincts that were sampled. In multistage sampling, you could introduce stratification. For example, you could first stratify

> *Multistage random sampling* is used in large-scale studies.

---

[3] If your purpose was to compare males with females, then it would be acceptable to draw the same number of each and compare averages or percentages for the two samples.

[4] Of course, if your hunch that males and females differ in their opinion was wrong, the use of stratification would be of no benefit, but it would not introduce any additional errors beyond the sampling errors created at random.

the counties into rural, suburban, and urban and then separately draw counties at random from these three types of counties — ensuring that all three types of counties are included.

A technique that is often useful is *cluster sampling* if it is done at random. To use cluster sampling, all members must belong to a cluster (i.e., an existing group). For example, all Boy Scouts belong to a troop; in most high schools, all students belong to a homeroom; etc. Unlike simple random sampling in which individuals are drawn, in cluster sampling, *clusters* are drawn. To conduct a survey of Boy Scouts, for example, one could draw a random sample of troops, contact the leaders of the selected troops, and ask them to administer the questionnaires. The advantages are obvious; there are fewer people to contact (only the leaders), and the degree of cooperation is likely to be greater if a leader asks the Scouts to participate.[5] There is a disadvantage, however, which results from the fact that clusters often are homogeneous in some way. This disadvantage is illustrated best by example. Suppose that you drew 10 troops (i.e., clusters) at random and nine of them, by chance, were in major urban areas. Scouts in major urban areas may have different attitudes and skills than those in rural areas. Even though there might be about 20 Scouts in each troop, yielding responses from about 200 Scouts, the number 200 is misleading because the sample size is 10 and *not* 200. If you had drawn 200 Boy Scouts by simple random sampling, it is very unlikely that you would get such a disproportionate number of Scouts from urban areas, and the sample size would be 200. Furthermore, had you used stratified random sampling and stratified on the basis of geographical area, you could have physically prevented such an error. Thus, when using cluster sampling, it is desirable to use a large number of clusters to overcome the disadvantage.

Suppose that we have wisely decided to use random sampling. How large should our sample be? This depends on a variety of factors, but here are a few generalizations that guide scientists:

1. The larger the sample, the better — but increasing sample size produces diminishing returns. For example, using a sample of

In cluster sampling, existing groups of subjects are drawn.

Increasing sample size produces diminishing returns.

---

[5]Keep in mind that for ethical and legal reasons, informed consent of the subjects or guardians of minors should be obtained before conducting most studies.

200 instead of 100 has a much greater effect on reducing sampling errors than using a sample of 3,100 instead of 3,000. In concrete terms, this means that an extra 100 subjects added to a small sample has a much greater effect on precision than adding 100 subjects to a large sample. In fact, in many national surveys, only 1,500 to 2,000 carefully selected subjects yield highly accurate results.

2. When there is little variability in a population, even a small sample may yield highly accurate results. For example, if you take a random sample of eggs that have been graded as "extra large" and weigh them, you will probably find only a small amount of variation among them. For this population, a small random sample should yield an accurate estimate of the average weight of "extra large eggs."

Small samples are adequate when there is little variability in the population.

3. When there is much variability in a population, small samples may produce data with much error. Suppose you wanted to estimate the math achievement of sixth-graders in a very large metropolitan school district and drew a random sample of only 100 students. Because there is likely to be tremendous variation in math ability across a large school district, a sample of 100, even though it is random, could be very misleading. By chance, for example, you may obtain a disproportionately large number of high achievers. Using a much larger sample would greatly reduce this possibility.

When there is much variability in a population, small samples may produce large errors.

4. When studying a rare phenomenon, large samples are usually required. For example, suppose you wanted to estimate the percentage of college students who are HIV-positive. If you draw a sample of only about 100, you probably would find no cases because the disease is relatively rare and is unlikely to be evident in such a small sample; thus, you might mistakenly conclude that no college students are HIV-positive. By using a sample of many thousands, you could get an accurate estimate of the small percentage who are positive.

Large samples are usually needed when studying a rare phenomenon.

It is important to remember that using a large sample does not correct for a bias. For example, if you are homeless and you ask many hundreds of your homeless friends how they feel about the government giving a thousand-dollar grant to each homeless person, you may mistakenly misjudge the thoughts of the general public on this subject. Even if you traveled all around the country and asked thousands of homeless people you encountered about this issue, you probably would be just as much in error if you were trying to predict the general public's attitude. This illustrates that in general, it is better to use a small, unbiased sample than a large, biased sample. Using a random process to select subjects eliminates bias.

It is better to use a small, unbiased sample than a large, biased sample.

Exercise for Section 15

Note: If any of your answers include the term *random sampling*, precede it with one of these adjectives: *simple*, *stratified*, or *multistage*.

Factual Questions

1. If you put the names of all members of a population on slips of paper, mix them, and draw some, what type of sampling are you using?

2. If there are 60 members of a population and you give them all number names starting with 01, what are the number names of the *first two subjects selected* if you select a sample starting at the beginning of the third line of the Table of Random Numbers in Appendix C?

3. If there are 500 members of a population and you give them all number names starting with 001, what are the number names of the *first two subjects selected* if you select a sample starting at the beginning of the fourth line of the Table of Random Numbers in Appendix C?

4. Suppose you draw at random the names of 5% of the registered voters separately from each county in a state. What type of sampling are you using?

5. Is stratification designed to *eliminate* or *reduce* sampling errors?

6. Suppose you draw 12 of the homerooms in a school at random and administer a questionnaire to all students in the selected homerooms. What type of sampling are you using?

7. Which of the following produces a greater reduction in sampling errors?
   A. increasing the size of a sample from 200 to 300
   B. increasing the size of a sample from 800 to 900

8. For which type of population should you use a larger sample?
   A. one with little variability
   B. one with much variability

9. If you are studying a rare phenomenon, should you use a large or small sample?

10. If you were forced to choose between the following, which should you usually select?
    A. a very large, biased sample
    B. a relatively small, unbiased sample

Question for Discussion

11. Suppose you want to conduct a survey of a sample of the students registered at your college or university. Briefly describe how you would select the sample.

Section 16

Standard Error of the Mean

Suppose that there is a large population with a mean of 100.00 and a standard deviation of 16.00 on a standardized test. Further, suppose that we do not have this information but wish to estimate the mean and standard deviation by testing a sample. When we draw a random sample and administer the test to just the sample, will we correctly estimate the population mean as 100.00? The answer is probably not. Remember that random sampling introduces random or chance errors — known as *sampling errors*.

Random (chance) errors are known simply as *sampling errors*.

At first, this situation may seem rather hopeless, but we have the advantage of using an unbiased, random sample — meaning that no factor(s) are systematically pushing our estimate in the wrong direction (i.e., there is no bias). In such a situation, using a large sample increases the odds that we are correct or, at least, not likely to make a large error.[1]

We also have the advantage of the *central limit theorem*. To understand this theorem, you must first understand the sampling distribution of means. Suppose that we drew not just one sample but many samples at random from the same population. That is, we drew a sample of 60, tested the subjects, and computed the mean; then drew another sample of 60, tested the subjects, and computed the mean; then drew another sample of 60, tested the subjects, and computed the mean; etc. We would then have a very large number of means — known as the *sampling distribution of means*. The central limit theorem says that the distribution of these means is normal in shape. The normal shape will emerge even if the underlying distribution is skewed, provided that the sample size is reasonably large (about 60 or more). The mean of an indefinitely large sampling distribution of means will equal the population mean. The standard deviation of the sampling distribution is known as the *standard error of the mean* (SE_M). Keep in mind that the means vary from each other only because of chance errors created by random sampling. That

According to the *central limit theorem*, the *sampling distribution of means* is normal.

The standard deviation of the sampling distribution is known as the *standard error of the mean* (SE_M).

---

[1]Review Sections 3 and 15 on sampling.

is, we are drawing random samples from the same population and administering the same test over and over, so all of the means should have the same value except for the effects of random errors. Therefore, there is variation among the means only because of sampling errors. Thus, once again, the *standard deviation of the sampling distribution* is known as the *standard error of the mean*.

In practice, we usually draw a single sample, test it, and calculate its mean and standard deviation. Therefore, we are not certain of the value of the population mean nor do we know the value of the standard error of the mean that we would obtain if we had sampled repeatedly. Fortunately, we do know two very useful things:

(1) The larger the sample, the smaller the standard error of the mean.

> The larger the sample, the smaller the standard error of the mean.

(2) The less the variability in a population, the smaller the standard error of the mean.

> The less the variability in a population, the smaller the standard error of the mean.

For example, consider a population in which there is no variability — that is, in which all subjects are identical. In this case, the standard error of the mean (i.e., the standard deviation of the sampling distributions of means) equals 0.00 (i.e., all the means will be identical and their standard deviation will be zero). In practice, we cannot be certain how much variability there is in a population from which we have only sampled. However, we can use the standard deviation of the sample that we have drawn as an estimate of the amount of variability in the population; for example, if we observed a very small standard deviation for a random sample, it would be reasonable to guess that the population has relatively little variation.

Given these two facts and some statistical theory that is not covered here, statisticians have developed a formula for estimating the standard error of the mean (SE_M) based on only the information we have about a given random sample from a population. Example 1 shows how it is sometimes reported after it has been calculated.

> For a given sample, the standard error of the mean is estimated with a formula.

Example 1:

You might read this statement in a research report: "m = 75.00, s = 16.00, n = 64, and SE_M = 2.00."

The SE_M in Example 1 is an estimate of a *margin of error* that we should keep in mind when interpreting the sample mean of 75.00. Also, keep in mind that the *standard error of the mean* is an estimate of the *standard deviation of the sampling distribution of the means*, which is normal in shape when the sample size is relatively large. You may recall from your study of the standard deviation that about 68% of the cases lie within one standard deviation unit of the mean. Thus, we would expect about 68% of all sample means to lie within 2.00 points of the true (or population) mean. If we use the sample mean of 75.00, which was actually obtained, as an estimate of the population mean based on a random sample, we could estimate that odds are 68 out of 100 that the population mean lies between 73.00 (75.00 - 2.00 = 73.00) and 77.00 (75.00 + 2.00). The values of 73.00 and 77.00 are known as the *limits of the 68% confidence interval for the mean*. That is, we have about 68% confidence that the true mean lies between 73.00 and 77.00.

The standard error of the mean is a margin of error.

We expect about 68% of all sample means to lie within one standard error of the mean.

By adding SE_M to the mean and subtracting SE_M from the mean, we obtain the limits of the 68% confidence interval for the mean.

Note that it is customary to report the sample size (n), mean (m), and standard deviation (s) when reporting the standard error of the mean.

Authors sometimes do the arithmetic for us and provide a statement like the one in Example 2, using *C.I.* as an abbreviation for *confidence interval*.

Example 2:

You might read this statement in a research report: "m = 75.00, s = 16.00, n = 64, and 68% C.I. = 73.00 — 77.00."

Scientists frequently build 95% or 99% confidence intervals and report one of them instead of a 68% interval.[2] When there are several groups, the statistics are often reported in a table such as the one shown in Example 3.

95% and 99% confidence intervals are often reported.

---

[2] The formulas for building 95% and 99% confidence intervals are beyond the scope of this book, but the results are illustrated in Example 3.

Example 3:

Table 16.1 Selected statistics

	n	*m*	*s*	*95% C.I.*
Group A	128	75.00	16.00	72.23–77.77
Group B	200	75.00	10.00	73.61–76.39

The 95% confidence intervals in Example 3 tell us the range of score values in which we can have 95% confidence that the true (i.e., population) mean lies.[3] Let's consider Group A; we can have 95% confidence that the true mean lies between 72.23 and 77.77.[4] Notice that we are not certain where the true mean lies because the 128 subjects are just a random sample of a population.

It should be obvious that a small confidence interval is desirable because it indicates that the sample mean is probably close to the true mean. Of the two variables that affect the size of the standard error of the mean — the sample size and the variability of the sample — we often have direct control of the sample size. By using reasonably large samples, we can minimize the standard error of the mean and obtain confidence intervals that are reasonably small.

It's important to keep in mind that the confidence limits are only valid if we analyze the results obtained with unbiased (random) sampling. Each bias has its own unique and usually unknown effects on the results, and there are no generalizable techniques for estimating the amount of error created by them.

A 95% C.I. tells us the range of values in which we can have 95% confidence that the true mean lies.

Using reasonably large samples helps keep the standard error of the mean and confidence intervals reasonably small.

Confidence intervals help us interpret means that are subject to random errors; they cannot take bias into account.

---

[3]Notice that the 95% C.I. for Group B is smaller than the one for Group A. This is because Group B has a larger *n* and a smaller *s*.
[4]Put another way, if we drew 100 samples of 128 subjects and constructed 95% confidence intervals for all 100 samples, about 95 of the 100 confidence intervals would include the true mean.

Exercise for Section 16

Factual Questions

1. *Sampling errors* refers to what kind of errors?

2. Very briefly state the central limit theorem.

3. What is the name of the standard deviation of the sampling distribution of means?

4. If you increase the size of a sample, what happens to the size of the standard error of the mean?

5. If there is no variation in a population, what is the value of the standard error of the mean?

6. Suppose a scientist reported the mean and standard error of the mean. How should you calculate the limits of the 68% confidence interval for the mean?

7. Suppose that you read that the mean equals 80.00 and the limits of the 95% C.I. are 75.00 and 85.00. Briefly explain what the limits tell you.

8. Do confidence limits help us interpret errors due to bias in sampling?

Notes:

Section 17

Introduction to the Null Hypothesis

Suppose that we draw a random sample of first-grade girls and a random sample of first-grade boys from a large school district in order to estimate the average reading achievement of both groups on a standardized test. Let's suppose we obtain these means:

Girls   Boys
$m = 50.00$ $m = 46.00$

This result suggests that girls, on the average, have higher achievement in reading. But do they really? Remember that we have tested only random samples of the boys and girls. It is possible that the difference we obtained is due only to the errors created by random sampling, which are known as *sampling errors*. In other words, it is possible that the population mean for boys and the population mean for girls are identical, and we found a difference between the means of the two randomly selected samples only because of the chance errors associated with random sampling. This possibility is known as the *null hypothesis*. For the difference between two means, it says that:

The true difference between the means (in the population) is zero.

This statement can also be expressed with symbols, as follows:[1]

$H_0: \mu_1 - \mu_2 = 0$
    Where:

        H_0 is the symbol for the null hypothesis.
        μ_1 is the symbol for the *population* mean for one group.

We may obtain a difference between two means only because of sampling errors.

For two means, the *null hypothesis* says that the true difference between two sample means is zero.

---

[1]See Appendix D for a different expression of the null hypothesis.

μ_2 is the symbol for the *population* mean for the other group

Another way to state the null hypothesis is:

There is no true difference between the means.

There are alternative ways to express the null hypothesis.

The null hypothesis may also be stated in the positive as follows:

The observed difference between the means was created by sampling error.

Most scientists are searching for differences among people and for explanations for the differences that they find. Therefore, most scientists do not undertake their studies in the hope of confirming the null hypothesis. Yet, once they have sampled at random, they are *stuck* with the null hypothesis as a possible explanation for any observed differences. They may also have their own personal hypothesis (i.e., their *research hypothesis*), which is often inconsistent with the null hypothesis. One possibility is that they believe that the average reading achievement of girls is higher than that of boys. This is known as a *directional hypothesis* because it states that one particular group's average is higher than the other. Expressed as symbols, this hypothesis is:

A scientist's personal hypothesis is known as the *research hypothesis*.

A *directional research hypothesis* states that one particular group is higher than the other.

$H_1: \mu_1 > \mu_2$
        Where:

        H_1 is the symbol for an *alternative hypothesis* (i.e., an alternative to the null hypothesis), which in this case is a *directional research hypothesis*.

        μ_1 is the symbol for the population mean for the group hypothesized to have a higher mean (in this case, the girls).

        μ_2 is the symbol for the population mean for the other group (in this case, the boys).

Another scientist may hold a *nondirectional hypothesis* as his or her *research hypothesis*. That is, he or she believes that there is a difference between boys' and girls' reading achievement, but that there is insufficient information to hypothesize as to which group is higher. In other words, the scientist is hypothesizing that there is a difference — that the two groups are not equal—but is not willing to speculate in advance on the direction of the difference. This is how to state a nondirectional research hypothesis in symbols:

$H_1: \mu_1 \neq \mu_2$
 Where:

> H_1 is the symbol for an *alternative hypothesis* (i.e., an alternative to the null hypothesis), which in this case is a *nondirectional research hypothesis*.
> μ_1 is the symbol for the population mean for one group.
> μ_2 is the symbol for the population mean for the other group.

Because it's easy to get lost the first time, let's reconsider the possibilities. Scientists conduct research in which two means are compared because they personally believe in one of three *research hypotheses*:

(1) There is no difference. (This research hypothesis is consistent with the null hypothesis, but is not frequently held.)

(2) One specific group is higher than the other. (This research hypothesis is the most frequently held and is inconsistent with the null hypothesis. It is a *directional* hypothesis.)

(3) There is a difference between the two groups in an unspecified direction. (This research hypothesis is not frequently held. It is inconsistent with the null hypothesis and is a *nondirectional* hypothesis.)

Remember, whichever hypothesis a scientist believes is true at the onset of a study is his or her *research hypothesis*. As you can see above, the

A *nondirectional research hypothesis* states only that there is a difference but does not indicate which group is higher.

Research hypotheses are usually inconsistent with *null hypotheses*.

most frequently held is number 2, a *directional research hypothesis*. Let's suppose that the two means we considered at the beginning of this section (i.e., $m = 50.00$ for girls and $m = 46.00$ for boys) were obtained by a scientist who started with the directional research hypothesis that girls achieve a higher mean in reading than boys. Clearly, the observed means support the research hypothesis, but is the scientist finished? Obviously not — because he or she has two possible explanations for the observed difference:

(1) Girls have higher achievement in reading than boys. (This is the research hypothesis.)

(2) The observed difference in the sample is the result of the effects of random sampling; therefore, there is no true difference. (This is the null hypothesis.)

> The *null hypothesis* is a possible explanation for an observed difference when there are *sampling errors*.

Notice that the scientist is at least temporarily *stuck* with the null hypothesis. Like it or not, because only a random sample was studied, he or she may be observing a difference that is the result of sampling errors. Thus, the null hypothesis is a possible explanation for the difference. If the scientist stops at this point, he or she has two explanations for a single difference of four points. This is hardly a definitive result. This scientist obviously should try to rule out the null hypothesis, leaving only the original research hypothesis. With the aid of inferential statistics, the scientist may test the null hypothesis. As a result of the inferential test, he or she may be able to rule it out.

The rest of this book deals with the null hypothesis and tests of it. In the next section, we will examine the logic of the null hypothesis in more detail.

Exercise for Section 17

Factual Questions

1. For the difference between two sample means, the null hypothesis says that the true difference in the population equals what?

2. What a scientist believes she will find before undertaking a study is known as what?

3. If a scientist believes that Group A will have a higher mean than Group B, is her research hypothesis *directional* or *nondirectional*?

4. According to this section, are *directional* or *nondirectional* research hypotheses most commonly held by scientists?

5. An investigator has studied *all* of the girls and boys in a population. She has found a difference between the mean for boys and the mean for girls. Is the null hypothesis a possible explanation for the difference? Explain.

6. What is the symbol for the null hypothesis?

7. What is the symbol for an alternative hypothesis?

8. For what does the symbol μ stand?

9. What term is defined as "errors created by random sampling"?

Notes:

Section 18

Decisions About the
Null Hypothesis

In the previous section, you learned that the null hypothesis states that there is no true difference between two means—that in the population, the difference is zero.[1] In other words, a difference between means was obtained only because of sampling errors created by random sampling.

An inferential test of a null hypothesis yields, as its final result, a *probability that the null hypothesis is true*. The symbol for probability is a lowercase *p*. Thus, if we find that the probability that the null hypothesis is true in a given study is less than 5 in 100, this result would be expressed as $p < .05$. How should this be interpreted? What does it tell us about the null hypothesis? Quite simply, it tells us that it is *unlikely* that the null hypothesis is true. If it is unlikely to be true, what should we conclude about it? We should conclude that it is probably not true.

It's important to understand the point of the previous paragraph, so let's consider an analogy. Suppose the weather reporter reports that the probability of rain tomorrow is less than 5 in 100. What should we conclude? First, we know that there is some chance of rain—but it is very small. Because of its low probability, most people would conclude that it probably will not rain and not make any special preparations for it. By not making any special preparations, they are acting as though it will not rain; for all practical purposes, they have rejected the hypothesis that it will rain tomorrow.

There is always some probability that the null hypothesis is true— so if we wait for certainty, we will never be able to make a decision. Thus, statisticians and applied researchers have settled on the .05 level as the level at which it is appropriate to reject the null hypothesis.[2] When

> An inferential test of a null hypothesis yields a *probability (p) that the null hypothesis is true.*

> When *p* is less than 5 in 100 that something is true, it is conventional to regard it as unlikely to be true.

> We can never be certain about the truth of the null hypothesis.

---

[1] See Appendix D for a different expression of the null hypothesis.
[2] The probability at which we are willing to reject the null hypothesis is known as the *alpha* level.

we use the .05 level, we are, in effect, willing to be wrong 5 times in 100 in rejecting the null hypothesis. Consider the rain analogy. If we make no special preparations for rain 100 days for which the probability of rain is .05, it will probably rain 5 of those 100 days. Thus, in rejecting the null hypothesis, we are taking a calculated risk that we might be wrong. This type of error is known as a *Type I error* — the error of rejecting the null hypothesis when it is correct. On those 5 days in 100 when you are caught in the rain without your rain gear, you will get wet because you made Type I errors.

A *Type I error* is the error of rejecting the null hypothesis when it is true.

In review, when the probability is low that the null hypothesis is correct, we reject the null hypothesis. A synonym for rejecting the null hypothesis is declaring a result to be *statistically significant*. In academic journals, you will find statements such as this: *The difference between the means is statistically significant.* This statement indicates that the authors have rejected the null hypothesis.

A synonym for *rejecting the null hypothesis* is declaring a result to be *statistically significant.*

In journals, you will frequently find *p* values of less than .05 reported.[3] The most common are $p < .01$ (less than 1 in 100) and $p < .001$ (less than 1 in 1,000). When a result is statistically significant at these levels, investigators can be more confident that they are making the right decision in rejecting the null hypothesis than they could be by using the .05 level. Clearly, if there is only 1 chance in 1,000 that something is true, it is less likely that it is true than if there are 5 chances in 100 that it is true. For this reason, the .01 level is a *higher* level of significance than the .05 level, and the .001 level is a *higher* level of significance than the .01 level. Let's review:

The lower the probability, the higher the level of significance. For example, $p < .01$ is a higher level of significance than $p < .05$.

.06+ level: *not* significant, do *not* reject the null hypothesis.

.05 level: significant, reject the null hypothesis.

.01 level: more significant, reject the null hypothesis with more confidence than at the .05 level.

.001 level: highly significant, reject the null hypothesis with even more confidence than at the .01 or .05 levels.

---

[3]Note that if *p* exactly equals .05, statistical significance is declared at the .05 level.

So what probability level should be used? Remember that most investigators are looking for significant differences (or relationships). Thus, they are most likely to use the .05 level because this is the easiest to achieve.[4]

Should you decide to use some level other than .05, you should decide that in advance of examining the data. Keep in mind though that when you require a lower probability before rejecting the null hypothesis (e.g., .01 instead of .05), you are increasing the odds that you will make a *Type II error*.[5] A Type II error is the error of failing to reject the null hypothesis when it is false. This type of error can have serious consequences. Suppose a drug company has developed a new drug for a serious disease. Suppose that, in reality, the new drug is effective. If, however, the null hypothesis is not rejected because the drug company selected a level of significance that is too high, the results of the study will have to be described as insignificant, and the drug may not receive government approval.

> A *Type II error* is the error of failing to reject the null hypothesis when it is false.

In review, there are two types of errors that may be made when making a decision about the null hypothesis:

Type I error: reject the null hypothesis when, in reality, it is true.

Type II error: fail to reject the null hypothesis when, in reality, it is false.

At first, this might seem frustrating — we've done our best, but we are still faced with the possibility of errors. While this is true, we will be making informed decisions, in the face of uncertainty, by using probabilities to our advantage. Either decision we make about the null hypothesis (reject or fail to reject) may be wrong, but by using inferential statistics to make the decisions, we can report to our readers the probability that we have made a *Type I error* (indicated by the p value we report). By reporting the probability level that we used, readers will be

> We are making informed decisions using probabilities in the light of uncertainty.

---

[4]However, if they find that their result is significant at the .01 or .001 levels, they will report it at these levels for the readers' information. On the other hand, if it had reached only the .05 level, they still would have reported it as significant in many cases. The next section provides more information on reporting the results of an inferential test.
[5]The probability of this type of error is known as *beta*.

informed of the odds that we were incorrect when we decided to reject the null hypothesis.

In this section and the previous section, I have emphasized comparing means because this is one of the most common procedures in statistical literature. The following four sections provide more information on statistical tests of significance for means. In addition, frequencies are often compared; this type of comparison is discussed in the last section of the book.[6]

Exercise for Section 18

Factual Questions

1. What is the name of the error of rejecting the null hypothesis when it is true?

2. If a difference is declared to be statistically significant, what decision is being made about the null hypothesis?

3. Is the .05 level or the .01 level a higher level of significance?

4. Is the .01 level or the .001 level a higher level of significance?

5. Is a difference usually regarded as *statistically significant* or *statistically insignificant* when $p < .05$?

---

[6]All other descriptive statistics may be compared for significance using inferential statistics. For example, we can compare medians, standard deviations, and correlation coefficients. Although the mathematical procedures and some of the theory are different, in any such comparisons, there is a null hypothesis that attributes differences to sampling errors. Decisions about the null hypotheses for other descriptive statistics are made using the same probability levels discussed in this section.

6. Is a difference usually regarded as *statistically significant* or *statistically insignificant* when $p < .001$?

7. Is a difference usually regarded as *statistically significant* or *statistically insignificant* when $p > .05$?

8. What is the name of the error of failing to reject the null hypothesis when it is, in reality, false?

9. Is it possible to reject the null hypothesis with certainty?

Notes:

Section 19

Introduction to the *t* Test

In this section, we will consider a frequently encountered problem: how to compare the means of two samples for statistical significance. Let's consider two examples:

This section deals with how to compare two sample means for significance.

Example 1:

An investigator wished to determine whether there are differences between men and women voters in their attitudes toward welfare. Samples of men and women were drawn at random and administered an attitude scale to obtain a score for each subject. Means for the two samples were computed. Women had a mean of 40.00 (on a scale from 0 to 50, where 50 is the most favorable). Men had a mean of 35.00. The researcher wishes to determine whether there is a significant difference between men and women. What accounts for the 5-point difference? One possible explanation is the *null hypothesis*, which states that there is no true difference between men and women — that the observed difference is due to sampling errors created by random sampling.

Example 1 illustrates that two means may be obtained from a *survey* —a descriptive study in which subjects are assessed in order to describe their traits at a given point in time.

Surveys sometimes yield two sample means that need to be tested for significance.

Example 2:

A random sample of kittens is fed a vitamin supplement from birth to see if the supplement increases their visual acuity. Another random sample is fed a placebo that looks like the supplement but contains no vitamins. At the end of the study, both samples are tested for visual acuity and an average acuity score is calculated for each sample. Those that took the supplement scored 4 points higher than the control group. What accounts for the 4-point difference? One possible explanation is the *null hypothesis*, which states that there is no true difference between the two samples of kittens—that the observed difference is due to sampling errors created by random sampling.

Example 2 illustrates that two means may be obtained from an *experiment*—a study in which treatments are given in order to observe for their effects.

Experiments sometimes yield two sample means that need to be tested for significance.

Surveys and experiments are very frequently conducted, and they often yield two means each, so you can see how important it is to be able to test the null hypothesis for the difference between two sample means.[1]

About a hundred years ago, a statistician named William Gosset developed the *t test* for exactly the situations we are considering. As a test of the null hypothesis, it yields a probability that a given null hypothesis is correct. When the probability that it is correct is low—say .05 or 5% or less—we usually reject the null hypothesis.

When the *t test* yields a low probability that a null hypothesis is correct, we reject the null hypothesis.

The computational procedures for conducting *t* tests are beyond the scope of this book. However, we will consider what makes the *t* test work. In other words, what leads the *t* test to give us a low probability that the null hypothesis is correct? Here are the three basic factors:

(1) The larger the samples, the less likely the difference between two means was created by sampling errors. You probably already intuitively knew that larger samples have less sampling error than smaller ones. Thus, when large samples are used, the *t* test is more likely to yield a probability low enough to allow us to reject the null hypothesis than when small samples are used.

The larger the sample, the more likely the null hypothesis will be rejected.

(2) The larger the observed difference between the two means, the less likely that the difference was created by sampling errors. Random sampling tends to create many small differences and few large ones. Thus, when large differences between means are obtained, the *t* test

The larger the observed difference between two means, the more likely the null hypothesis will be rejected.

---

[1]Other types of studies also yield two sample means that are to be compared.

is more likely to yield a probability low enough to allow us to reject the null hypothesis than when small differences are obtained.

(3) The smaller the variance among the subjects, the less likely that the difference between two means was created by sampling errors and the more likely the null hypothesis will be rejected. To understand this, consider a population in which everyone is identical—they all look alike, think alike, and speak and act in unison. How many do you have to sample to get a good sample? Only one since they are all the same. Thus, when there is no variation among subjects, it is not possible to have sampling errors. If there are no sampling errors, the null hypothesis should be rejected. As the variation increases, sampling errors are more and more likely.[2]

The smaller the variance, the more likely the null hypothesis will be rejected.

There are two types of *t* tests. One is for *independent data* (sometimes called *uncorrelated data*), and one is for *dependent data* (sometimes called *correlated data*). Examples 1 and 2 have independent data. Example 3 describes a study with dependent data:

The first two examples illustrate independent data.

Example 3:
In a study of visual acuity, same-sex siblings (two brothers or two sisters) were identified for a study. For each pair of siblings, a coin was tossed to determine which one received a vitamin supplement and which received a placebo. Thus, for each subject in the experimental group, there is a same-sex sibling in the control group.

Example 3 illustrates dependent data.

The means that result from the study in Example 3 are subject to less error than the means from Example 2. Remember that in Example 2 there was no matching or pairing of subjects before assignment to conditions. In Example 3, the matching of subjects assures us that the two groups are more similar than if just any two independent samples were used. To the extent that genetics and gender are associated with visual acuity, the two groups in Example 3 will be more similar at the onset of the experiment than the two groups in Example 2. The *t* test for dependent data

Dependent data may have less sampling error.

---

[2] In the types of studies we are considering, we do not know the population standard deviation, which would indicate the amount of variation. The *t* test uses the standard deviations of the samples to estimate the variation of the population.

takes this possible reduction of error into account.[3] Thus, it is important to select the appropriate *t* test. Scientists will often mention in their reports whether they conducted *t* tests for independent or for dependent data.

The next section illustrates how to interpret reports of *t* tests.

Exercise for Section 19

Factual Questions

1. What can be compared for significance using a *t* test?

2. How is a survey different from an experiment?

3. If a *t* test yields a low probability, what decision is usually made about the null hypothesis?

4. The larger the sample, the
    A. more likely the null hypothesis will be rejected.
    B. less likely the null hypothesis will be rejected.

5. The smaller the observed difference between two means, the
    A. more likely the null hypothesis will be rejected.
    B. less likely the null hypothesis will be rejected.

6. The smaller the variance, the
    A. more likely the null hypothesis will be rejected.
    B. less likely the null hypothesis will be rejected.

---

[3]Ideally, we would like to conduct an experiment in which the two groups are initially *identical* in their visual acuity.

7. If subjects are first paired before being randomly assigned to experimental and control groups, are the resulting data *independent* or *dependent*?

8. Which type of data tend to have less sampling error?
   A. Independent.   B. Dependent.

Notes:

Section 20

Reporting the Results of *t* Tests

We are considering the use of the *t* test to test the difference between two sample means for significance.[1] Obviously, the values of the means should be reported before reporting the results of the test on them. In addition, the values of the standard deviations and the number of cases in each group should be reported. This may be done within the context of a sentence or in a table. Table 20.1 shows a typical table.

The means, standard deviations, and numbers of cases should be reported before reporting the results of a *t* test.

Table 20.1 Means and standard deviations

	m	*s*	*n*
Group A	2.50	1.87	6
Group B	6.00	1.89	6

The samples that formed Groups A and B were drawn at random. The null hypothesis states that the 3.5 point difference (6.00 - 2.50 = 3.50) between the means of 2.50 and 6.00 are the result of sampling errors (i.e., errors resulting from random sampling) and that the true difference in the population is zero.[2]

The result of a significant *t* test can be described in several ways. Here are some examples for the results in Table 20.1:

Example 1:
The difference between the means is statistically significant ($t = 3.22$, $df = 10$, $p < .01$).

Example 1 shows one way to report the result of a significant *t* test.

---

[1] The *t* test can also be used to test the significance of the difference between two Pearson *r*'s.
[2] This statement of the null hypothesis complements a nondirectional research hypothesis. See Sections 17 and 18.

The statement in Example 1 indicates to the sophisticated reader that the null hypothesis has been rejected because *statistically significant* is synonymous with *rejecting the null hypothesis*.

Statistically significant is synonymous with rejecting the null hypothesis.

Example 2:
The difference between the means is significant at the .01 level ($t = 3.22$, $df = 10$).

In Example 2, the author has used slightly different wording to indicate that significance was obtained at the .01 level. The phrase *significant at the .01* level tells us that p was equal to or less than .01. Thus, the null hypothesis was rejected.

The phrase *significant at the .01* level tells us that p was equal to or less than .01.

Example 3:
The null hypothesis was rejected at the .01 level ($t = 3.22$, $df = 10$).

From Example 3, we know that the difference is statistically significant because *rejecting the null hypothesis* is the same as *declaring statistical significance*.

Rejecting the null hypothesis is the same as declaring statistical significance.

Any of the three forms shown above is acceptable. Authors of journal articles tend to talk about differences as being either "statistically significant" or "statistically insignificant" and seldom mention the null hypothesis. In theses and dissertations, explicit references to the null hypothesis are more common.

When you use the word *significant* when reporting the results of significance tests, you should always modify it with the adjective *statistically*. This is because a result may be *statistically significant* but not of any *practical significance*. For example, suppose you found a statistically significant difference of 2 points in favor of a computer-assisted approach over a traditional lecture/textbook approach. While it is statistically significant, it may not be of practical significance if the school district has to invest sizable amounts of money to buy new hardware and software; in other words, the cost of the difference may be too great in light of the absolute size of the benefit.

A result may be *statistically significant* but not of any *practical significance*.

Let's consider how to report a difference that is not significant. Table 20.2 presents descriptive statistics. Examples 4 through 6 show some ways to express the results of the *t* test performed on the difference between the two means.

Table 20.2 Means and standard deviations

	m	*s*	*n*
Group X	8.14	2.19	7
Group Y	5.71	2.81	7

Example 4:

The difference between the means is not statistically significant ($t = 1.80$, $df = 12$, $p > .05$).

Examples 4, 5, and 6 show how the results of an insignificant *t* test may be reported.

The fact that *p* is *greater than* (>) .05 indicates that we should not reject the null hypothesis and not declare statistical significance.[3]

Example 5:

For the difference between the means, $t = 1.80$ ($df = 12$, *n.s.*).

The author of Example 5 has used the abbreviation *n.s.* to tell us that she has declared the difference to be not significant. Because we are not given a specific probability level, most readers will assume that it was not significant at the .05 level — the most liberal of the widely used levels.[4] Example 4 is preferable to Example 5 because Example 4 indicates the specific probability level in question.

The abbreviation *n.s.* means *not significant.*

It is best to indicate the specific probability level at which the null hypothesis was not rejected.

Example 6:

The null hypothesis for the difference between the means was not rejected at the .05 level ($t = 1.80$, $df = 12$).

---

[3]You may recall that the .05 level is the lowest level of significance in most sciences. See Section 18.
[4]Saying that a difference was *not significant at the .05 level* is synonymous with saying that *p is greater than .05* (i.e., $p > .05$).

While reading journal articles, theses, and dissertations, you will find variations in the exact words used to describe the results of *t* tests. The examples in this section show you some of the most widely used forms of expression.

Exercise for Section 20

Factual Questions

1. Which statistics should be reported before reporting the results of a *t* test?

2. Suppose you read this statement: "The difference between the means is statistically significant at the .05 level ($t = 2.333$, $df = 11$)." Should you conclude that the null hypothesis has been rejected?

3. Suppose you read this statement: "The null hypothesis was rejected ($t = 2.810$, $df = 40$, $p < .01$)." Should you conclude that the difference is statistically significant?

4. Suppose you read this statement: "For the difference between the means, $t = 2.111$ ($df = 5$, *n.s.*)." Should you conclude that the null hypothesis has been rejected?

5. For the statement in question 4, should you conclude that the difference is statistically significant?

6. What important type of information is missing in the statement in question 4?

Section 21

One-Way ANOVA

In Sections 19 and 20, you learned about the t test, which tests the null hypothesis regarding the difference between *two* means. A closely related test is *analysis of variance* (*ANOVA*), which is sometimes informally called the F test. ANOVA is used to test the difference(s) among *two or more* means.

First, ANOVA can be used to test the difference between two means. When this is done, the resulting *probability* will be the same as the probability that would have been obtained using a t test. However, the value of F will not be the same as the value of t.

ANOVA can also be used to test the differences among more than two means in a single test — which cannot be done with a t test. Consider Example 1:

Example 1:
A new drug for treating migraine headaches was tested on three groups selected at random. The first group received 250 milligrams, the second received 100 milligrams, and the third received a placebo (an inert substance). The average reported pain for the three groups (on a scale from 0 to 20, with 20 representing the most pain) was determined by calculating means. The means for the groups were:

> Group 1: $M =$ 1.78
> Group 2: $M =$ 3.98
> Group 3: $M =$ 12.88

As you can see in Example 1, there are three differences among means:

(1) the difference between Groups 1 and 2,

(2) the difference between Groups 1 and 3, and

ANOVA stands for *analysis of variance*; it is sometimes called the F test.

ANOVA is used to test the difference(s) among two or more means.

For the difference between two means, t and ANOVA yield the same probability.

(3) the difference between Groups 2 and 3.

Instead of running three separate *t* tests,[1] we can run a single ANOVA to test the significance of this *set of differences*. There are two ways the results of Example 1 can be reported. Example 2 shows one of these:

The *set of three differences* in Example 1 can be tested with a single ANOVA.

Example 2:

The difference among the means was statistically significant at the .01 level ($F = 58.769$, *df* = 2, 24).

Note that the method of reporting in Example 2 is similar to that for reporting the results of a *t* test.[2] This result tells us that there is a significant difference with $p < .01$. Thus, the null hypothesis may be rejected at the .01 level. The null hypothesis for this test says that the *set of three differences* was created at random. By rejecting the null hypothesis, we are rejecting the notion that *one or more* of the differences were created at random. Notice that the test does not tell us which of the three differences are responsible for the rejection of the null hypothesis. It could be that only one or two of the three differences was responsible for significance. Procedures for determining which individual differences are significant are beyond the scope of this book.

The method of reporting the results of an ANOVA is sometimes similar to the method for a *t* test.

When we reject the null hypothesis with ANOVA, we are rejecting the notion that *one or more* of the differences in the set were created at random.

Example 3 shows another way that the results of an ANOVA are commonly reported in journals. It is called an *ANOVA table*. In the table, you see the values of *F*, *df* , and *p* that were reported in Example 2. You also see the values of the *sum of squares* and *mean square* — these are intermediate values obtained in the calculation of *F*. (For example, if you divide the mean square of 315.592 by the mean square of 5.370, you will obtain *F*.) For the typical consumer of research, however, these values are of little interest.[3] The typical consumer is primarily interested in whether or not the null hypothesis has been rejected, which is indicated by the value of *p*.

An ANOVA table is sometimes used to report the results.

The *sum of squares* and *mean square* in an ANOVA table are of little interest to the typical consumer of research.

---

[1] It would be inappropriate to run three separate *t* tests without an adjustment in the standard probabilities for *t*.

[2] You probably noticed that there are two values reported as degrees of freedom for an ANOVA; you may recall that there is only one value for *df* reported for a *t* test.

[3] Those with advanced training in statistics can use these intermediate values to enhance their interpretation of the data.

Example 3:

Table 1: Analysis of Variance Table for the data in Example 1.

Source of Variation	*df*	Sum of Squares	Mean Square	*F*
Between groups	2	631.185	315.592	58.769*
Within groups	36	193.320	5.370	
Total	38	824.505		

*$p < .01$

Notice that the probability in Example 3 is given in a footnote, which is common. However, sometimes it will be given in the table and sometimes it will be given in the text that describes the table.

It is common to report the probability level in a footnote to an ANOVA table.

The technique we are considering can be generalized to a larger number of means. Consider Example 4.

Example 4:

Four methods of teaching computer literacy were used in an experiment, which resulted in four means. This produced these six differences:

1. The difference between Methods 1 and 2.
2. The difference between Methods 1 and 3.
3. The difference between Methods 1 and 4.
4. The difference between Methods 2 and 3.
5. The difference between Methods 2 and 4.
6. The difference between Methods 3 and 4.

When there are four means, there are six differences among pairs of means.

A single ANOVA can determine whether the null hypothesis for this entire set of six differences should be rejected. If the result is not significant, the researcher is done. If the result is significant, he or she may conduct additional tests to determine which specific difference(s) are significant.[4] While these additional tests are beyond the scope of this

A single ANOVA can test the entire set of six differences.

---

[4] A number of different tests, which do not always lead to the same conclusions, are available. Some that you may encounter are Tukey's *HSD* Test and Scheffé's Test.

book, you will be able to understand them because they all result in a probability level (*p*), which is used to determine significance.

The examples we have been considering are examples of what is known as a *one-way ANOVA* (also known as *single-factor ANOVA*). This term is derived from the fact that subjects were classified *one* way. In Example 1, they were classified only according to the drug group to which they were assigned. In Example 4, they were classified only according to the method of instruction to which they were exposed. In the next section, you will be introduced to a *two-way ANOVA* (also known as a *two-factor ANOVA*) in which each subject is classified in two ways such as (1) which drug group they were assigned to and (2) whether he/she is male or female. A two-way ANOVA permits us to answer questions that are potentially more interesting.

The examples in this section are examples of a *one-way ANOVA*, in which each subject is classified in only *one* way.

In a *two-way ANOVA*, each subject is classified in two ways.

Exercise for Section 21

Factual Questions

1. *ANOVA* stands for what words?

2. A single ANOVA can be used to test the difference(s) among how many means?

3. If the difference between a pair of means is tested with ANOVA, will the probability level be different than if the difference was tested with a *t* test?

4. Which statistic in an ANOVA table is of the greatest interest to the typical consumer of research?

5. Suppose you read this statement: "The difference between the means was not statistically significant at the .05 level ($F = 2.293$, $df = 12, 18$)." Should you conclude that the null hypothesis was rejected?

6. Suppose you read this statement: "The difference between the means was statistically significant at the .01 level ($F = 3.409$, $df = 14, 17$)." Should you conclude that the null hypothesis was rejected?

7. Suppose that the subjects were classified according to their grade level in order to test the differences among the means for the grade levels. Does this call for a *one-way ANOVA* or a *two-way ANOVA*?

8. Suppose that the subjects were classified according to their grade levels and their country of birth in order to study differences among means for both grade level and country of birth. Does this call for a *one-way ANOVA* or a *two-way ANOVA*?

Question for Discussion

9. Briefly describe a study you might wish to conduct in which you think it would be appropriate to conduct a one-way ANOVA.

Notes:

Section 22

Two-Way ANOVA

In a *two-way ANOVA* (also known as a *two-factor ANOVA*), subjects are classified in two ways. Consider Example 1, which illustrates a two-way ANOVA.

Example 1:

A random sample of welfare recipients was assigned to a new job training program. Another random sample was assigned to a conventional job training program. (*Note*: Which of the job training programs they were assigned to is one of the ways in which the subjects were classified.) Subjects were also classified according to whether or not they had a high school diploma. All of the subjects in each group found employment in the private sector at the end of their training. Their mean hourly wages are shown in this table:[1]

	Type of Program		
	Conventional	New	**Row Means**
H.S. Diploma	$M = \$8.88$	$M = \$8.75$	$M = \$8.82$
No H.S. Diploma	$M = \$4.56$	$M = \$8.80$	$M = \$6.68$
Column Means	$M = \$6.72$	$M = \$8.78$	

First, let's consider the column means of $6.72 (for the conventional program) and $8.78 (for the new program). These suggest that overall, the new program is superior to the conventional one. In other words, if we temporarily ignore whether subjects have a high school diploma, the new program seems superior to the conventional one. This difference

---

[1] Note that income in large populations is usually skewed, making the mean an inappropriate average (see Section 11); for these groups, assume that it was not skewed. Also note that the row means and column means were obtained by adding and dividing by two; this is appropriate only if the number of subjects in all cells is equal; if it is not, compute the row and column means using the original raw scores.

($8.78 - $6.72 = $2.06) suggests that there is what is called a *main effect*. A *main effect* is the result of comparing one of the ways in which the subjects were classified while temporarily ignoring the other way in which they were classified.

A *main effect* results from examining one way in which subjects were classified while ignoring the other.

Since the concept of *main effect* is important, let's look at it another way. You can see that the column mean of $6.72 is for all of those who had the conventional program regardless of whether they have a high school diploma. The column mean of $8.78 is for all of those who had the new program regardless of whether they have a high school diploma. Thus, by looking at the column means, we are considering only the effects of the type of program (and *not* the effects of a high school diploma). When you look at the effects of only one way in which the subjects were classified, this is called a *main effect*.

If there is a difference between the column means, this suggests one main effect.

Now, consider the row means of $8.82 (for those with a high school diploma) and $6.68 (for those with no high school diploma). This suggests that those with a diploma, on the average, have higher earnings than those without one. This is also a *main effect*. This main effect is for high school diploma while temporarily ignoring the type of training program.

If there is a difference between the row means, this suggests another main effect.

To this point, we have two findings that would be of interest to those studying welfare: (1) the new program seems to be superior to the conventional program in terms of hourly wages and (2) those with a high school diploma seem to have higher hourly wages. (*Note*: The term "seem" is being used because we have only studied random samples, and we do not know yet whether the differences are statistically significant.)

You may have already noticed that there is a third interesting finding: Those with a high school diploma earn about the same amount regardless of the program. This statement is based on these means for *those with a high school diploma* reproduced from the table shown above:

	Conventional Program	New Program
H.S. Diploma	$M = \$8.88$	$M = \$8.75$

However, those with no high school diploma appear to benefit more from the new program than the conventional one. This statement is based on these means for *those with no high school diploma:*

	Conventional Program	**New Program**
No H.S. Diploma	$M = \$4.56$	$M = \$8.80$

Suppose you were the researcher who conducted this study; you are now an expert on the subject and an administrator calls you for advice. She asks you, "Which program should we use? The conventional one or the new one?" You could, of course, tell her that there is a *main effect* for programs, which suggests that overall the new program is superior in terms of wages — but if you stopped there, your answer would be incomplete. A more complete answer is:

(1) For those with a diploma, the two programs are about equal in effectiveness. Thus, the choice of a program for them should probably hinge on other considerations such as the cost of the two programs.

(2) For those with no diploma, the new program is superior to the conventional one. Other things being equal, those without a diploma should be assigned to the new program.

Because you cannot give a complete answer about the two types of programs (one way in which the subjects were classified) without also referring to high school diplomas (the other way in which they were classified), we say there is an *interaction* between the two. How well the two programs work depends, in part, upon whether the subjects have high school diplomas.

Here is a simple way in which you can spot an interaction when there are only two rows of means: Subtract each mean in the second row from the mean in the first row. If the two differences are the same, there is no interaction. If they are different, there is an interaction. Here is how it works for the data in Example 1:

When you cannot give a complete discussion about one main effect without discussing the other main effect, you have an *interaction*.

You can spot an interaction by subtracting down the rows and comparing the differences.

	Type of Program	
	Conventional	New
H.S. Diploma	$M = \$8.88$	$M = \$8.75$
No H.S. Diploma	$M = \$4.56$	$M = \$8.80$
Difference	**$4.32**	**-$0.05**

Example 1 has two main effects and an interaction.

Because the two differences are *not* the same, there is an interaction.

Consider Example 2, in which there are no main effects but there is an interaction.

Example 2:

A random sample of subjects from a population of those suffering from a chronic illness were administered a new drug. Another random sample from the same population was administered a standard drug. Subjects were also classified as to whether they were male or female. At the end of the study, improvement was measured on a scale from 0 (for no improvement) to 10 (for complete recovery). These means were obtained:

	Drug		
	Standard	New	**Row Means**
Male	$M = 5.00$	$M = 7.00$	**$M = 6.00$**
Female	$M = 7.00$	$M = 5.00$	**$M = 6.00$**

Example 2 has no main effects but has an interaction.

The two column means in Example 2 are the same. Thus, if we temporarily ignore whether subjects are male or female, we would conclude that the two drugs are equally effective. To state it statistically, we would say that *there is no main effect for the drugs*.

The two row means are the same. Thus, if we temporarily ignore which drug was taken, we can conclude that males and females improved to the same extent. To state it statistically, we would say that *there is no main effect for gender*.

Of course, the interesting finding in Example 2 is the *interaction*. The standard drug works better for females and the new drug works better for males. Subtracting as we did in Example 1, we obtain the differences shown below. Because -2.00 is not equal to 2.00, there is an interaction.

	Drug	
	Standard	New
Male	$M = 5.00$	$M = 7.00$
Female	$M = 7.00$	$M = 5.00$
Difference	**-2.00**	**2.00**

Consider Example 3 below, in which there are two main effects but no interaction.

Example 3:

Random samples of high and low achievers were assigned to one of two types of reinforcement during math lessons. Achievement on a math test at the end of the experiment was the outcome variable. The mean scores on the test were:

Example 3 has two main effects but no interaction.

	Type of Reinforcement		
	Type A	Type B	**Row Means**
High achievers	$M = 50.00$	$M = 30.00$	$M = 40.00$
Low achievers	$M = 40.00$	$M = 20.00$	$M = 30.00$
Column Means	$M = 45.00$	$M = 25.00$	

In Example 3, there seems to be a main effect for type of reinforcement as indicated by the difference between the column means (45.00 and 25.00). Thus, ignoring achievement levels temporarily, Type A seems to be more effective than Type B.

There also seems to be a main effect for achievement level as indicated by the difference between the row means (40.00 and 30.00). Thus, ignoring the type of reinforcement, high achievers score higher on the math test than low achievers.

There is no interaction, as indicated by the differences, which are shown below:

	Type of Reinforcement	
	Type A	Type B
High achievers	$M = 50.00$	$M = 30.00$
Low achievers	$M = 40.00$	$M = 20.00$
Difference	**10.00**	**10.00**

What does this lack of an interaction tell us? That regardless of the type of reinforcement, high achievers are the same number of points higher than low achievers (i.e., 10 points). Put another way, regardless of whether students are high or low achievers, Type A reinforcement is better.[2]

We are not restricted to two categories for each classification variable. We could, for example, study three types of reinforcement and obtain means for the following cells and still consider the two main effects and the interaction.

You can use more than two categories on each variable and still use ANOVA.

	Type of Reinforcement			
	Type A	Type B	Type C	**Row Means**
High Achievers	$M =$	$M =$	$M =$	$M =$
Low Achievers	$M =$	$M =$	$M =$	$M =$
Column Means	$M =$	$M =$	$M =$	

In review, a two-way ANOVA examines two *main effects* and one *interaction*. Of course, because only random samples have been examined, the null hypothesis must be considered. For each of the main effects and for the interaction, the null hypothesis states that there is no *true* difference —that the observed differences were created by random

---

[2]The basis for this second statement is that if you subtract across the rows, you get the same difference for each row. Earlier, you were told to subtract down columns; however, subtracting across the rows works equally well in determining whether there is an interaction.

sampling errors. A two-way ANOVA will, therefore, test the two main effects and the interaction for significance. This is done by computing three values of F (one for each of the three null hypotheses) and determining the probability associated with each. Typically, if a probability is .05 or less, the null hypothesis is rejected and the main effect or interaction is declared to be statistically significant.

A two-way ANOVA will test the null hypotheses for the two main effects and the interaction.

The results of a two-way ANOVA are usually organized in a table. While the entries in such tables sometimes vary,[3] the most important entries are shown in Example 4.

Results of a two-way ANOVA are usually organized in a table.

Example 4: An ANOVA Table

Source	df	F	p
Achievement Level	2,125	3.25	.042
Type of Reinforcement	6,120	19.69	.001
Interaction (Ach. x Reinf.)	12,240	1.32	.210

The probabilities shown in Example 4 lead us to our interpretation. Both main effects (i.e., achievement level and type of reinforcement) are statistically significant because the values of p are both less than .05, the lowest commonly accepted level for rejecting the null hypothesis. The interaction, however, is not significant because p is greater than .05, and the null hypothesis regarding this interaction should not be rejected. Thus, three null hypotheses have been tested with a single two-way ANOVA.

The probabilities (p) determine whether we should reject the null hypotheses.

Exercise for Section 22

Factual Questions

For all factual questions, assume that there are equal numbers of subjects in each cell.

---

[3]It is also common to include sum of squares and mean squares in an ANOVA table. As mentioned in the previous section, these are of little interest to the typical consumer of research.

Questions 1 through 3 refer to this information:

Random samples of subjects with back pain and with headache pain were randomly assigned to two types of pain relievers. The means indicate the average amount of pain relief.

	Type of Pain Reliever		
	Type X	Type Y	Row Means
Back pain	$M = 15.00$	$M = 25.00$	$M =$
Headache pain	$M = 10.00$	$M = 20.00$	$M =$
Column Means	$M =$	$M =$	

1. Does there seem to be a main effect for type of pain?

2. Does there seem to be a main effect for type of pain reliever?

3. Does there seem to be an interaction?

Questions 4 through 6 refer to this information:

Two types of basketball instruction were tried with random samples of subjects who either had previous experience playing or did not have previous experience. The means indicate the proficiency at playing basketball at the end of treatment.

	Type of Instruction		
	New	Conventional	Row Means
Had previous experience	$M = 230.00$	$M = 200.00$	$M =$
Did not have previous experience	$M = 200.00$	$M = 230.00$	$M =$
Column Means	$M =$	$M =$	

4. Does there seem to be a main effect for type of instruction?

5. Does there seem to be a main effect for experience?

6. Does there seem to be an interaction?

Question for Discussion

7. Briefly describe a study you might wish to conduct in which it would be appropriate to conduct a two-way ANOVA.

Notes:

Section 23

Chi Square

Frequently, our research data are *nominal* (i.e., naming data such as subjects naming the political candidates for whom they plan to vote).[1] Such data do not directly permit the computation of means and standard deviations. Instead, we usually report the number of subjects who named each category (i.e., the frequency) and the corresponding proportions or percentages. Here's an example:

Example 1:

A random sample of 200 likely voters was drawn and asked which of two candidates for an elected office they plan to vote for. These data were obtained:

Candidate Smith   Candidate Doe

$n = 110$ (55.0%) $n = 90$ (45.0%)

The data in Example 1 suggest that Candidate Smith is leading. However, only a random sample of likely voters was surveyed. It is possible, for example, that the population of likely voters is evenly split, but that a difference of 10 percentage points was obtained because of the sampling errors associated with random sampling. For this possibility, there is a null hypothesis that says that there is no true difference in the population — that is, in the population, the likely voters are evenly split. We cannot use a *t* test or ANOVA to test this null hypothesis because we do not have means. The appropriate test for the data under consideration (i.e., frequencies or numbers of cases) is *chi square*,[2] whose symbol is χ^2. A chi square test for this data indicates that the probability that the null hypothesis is true is greater than 5 in 100 ($p > .05$). Thus, we cannot reject the null hypothesis; the difference is not statistically significant. In

For *nominal* data, we report frequencies and percentages instead of means and standard deviations.

To test differences among frequencies, we use *chi square*. Its symbol is χ^2.

---

[1] See Section 4 to review scales of measurement, including nominal.
[2] The tests on means (*t* and *F*) in earlier sections are based on the assumption that the underlying distributions are normal; they are examples of *parametric tests*. Since chi square is not based on such an assumption, it is an example of a *nonparametric* (or *distribution-free*) test.

concrete terms, Candidate Smith cannot rest easy because we cannot rule out sampling errors as an explanation for the difference in her favor.

Example 1 illustrates a *one-way chi square* (also known as a *goodness of fit chi square*). The subjects are classified in only one way — for whom they plan to vote. Example 2 illustrates a *two-way chi square* in which samples from two populations of voters were classified in terms of for whom they plan to vote.[3]

In a one-way chi square, subjects are classified in only one way.

Example 2:

A random sample of 200 male registered voters and a random sample of 200 female registered voters were drawn and asked which of two candidates for an elected office they plan to vote for. These data were obtained:

	Candidate Jones	Candidate Black
Males	$n = 80$	$n = 120$
Females	$n = 120$	$n = 80$

In a two-way chi square, subjects are classified in two ways.

Inspection of the data in Example 2 suggests that Jones is a stronger candidate among females and Black is a stronger candidate among males. If this pattern is true among all males and all females in the voting population, both candidates should take heed. For example, candidate Jones might consider ways to shore up her support among males without alienating the females while candidate Black might do the opposite. However, only a random sample was surveyed. Before taking action, the candidates should consider how likely it is that the observed differences in preferences between the two groups (males and females) were created by random sampling errors. Chi square is the appropriate test because we are examining differences in nominal data. For this example, the chi square test reveals that the probability is less than 1 in 1,000 ($p < .001$) that the null hypothesis of no difference between the two populations is true. Thus, it is very unlikely that this pattern of differences is due to sampling errors — with a high degree of confidence, the candidates can rule out chance as an explanation.

The data in Example 2 suggest a relationship between gender and preference for candidates.

---

[3]There are two types of two-way chi square tests. Example 2 illustrates a *chi square test of homogeneity*. This test involves two or more populations (in this example, males and females) and their opinion on one outcome variable (in this example, which candidate they plan to vote for). Example 3 illustrates a *chi square test of independence* in which one population is classified in two ways.

In Example 3, a sample from one population of subjects was asked two questions, each of which yielded nominal data.

Example 3:

A random sample of college students was asked whether they think that IQ tests measure innate intelligence and whether they had taken a tests and measurements course. These data resulted:

	Took Course	Did Not Take Course
Yes, Innate	$n = 20$	$n = 30$
No, Not Innate	$n = 40$	$n = 15$

The observed data suggest that those who did not take the course were more likely to perceive IQ tests as measuring innate intelligence (30 vs. 15) than those who took the course (20 vs. 40). In other words, there appears to be a relationship between whether subjects have taken the course and what they believe IQ tests measure. Once again, only a random sample was questioned. Therefore, it is possible that the observed relationship is not true in the population. That is, the null hypothesis asserts that there is no *true* relationship (in the population). A chi square test for these data produced this result:

$$\chi^2 = 11.455, \, df = 1, \, p < .001$$

Thus, the null hypothesis may be rejected with a high degree of confidence since the odds are less than 1 in 1,000 that it is a true hypothesis.[4] Example 4 shows a typical statement that might appear in a journal.

Example 4:

The relationship was statistically significant with those who took the course being more likely to believe that IQ tests do not measure innate intelligence ($\chi^2 = 11.455, \, df = 1, \, p < .001$).

The data in Example 3 suggest a relationship between whether subjects have taken the course and what they believe IQ tests measure.

The chi square test for Example 3 indicates that the null hypothesis should be rejected at the .001 level; the relationship is statistically significant.

---

[4]Notice that in our examples, the responses are independent. For instance, in Example 2, the gender of the respondents is not determined by their preference for a candidate. Also, each response is mutually exclusive. For example, we do not allow a subject to indicate that he/she is both male and female. Independence and mutually exclusive categories are assumptions underlying chi square.

Exercise for Section 23

Factual Questions

1. If you calculated the mean score for freshmen and the mean score for seniors on a grammar test and wished to compare the two means for statistical significance, would a chi square test be appropriate?

2. If you asked a random sample of subjects which of two types of rice they prefer and wished to compare the frequencies with an inferential test, would a chi square test be appropriate?

3. For examining relationships for nominal data, do we use a *one-way chi square* or a *two-way chi square*?

4. Suppose you read that "$\chi^2 = 4.111$, $df = 1$, $p < .05$." What decision should be made about the null hypothesis at the .05 level?

5. Suppose you read that "$\chi^2 = 7.418$, $df = 1$, $p < .01$." Is this statistically significant at the .01 level?

6. Suppose you read that "$\chi^2 = 2.824$, $df = 2$, $p > .05$." What decision should be made about the null hypothesis at the .05 level?

7. If as a result of a chi square test, p is found to be less than .001, the odds that the null hypothesis is correct are less than 1 in ___?

Appendix A

Computation of the Standard Deviation

The formula that defines the standard deviation is:

$$S = \sqrt{\frac{\Sigma x^2}{N}}$$

The lowercase letter x stands for the deviation of a score from the mean of its distribution. To obtain it, first calculate the mean (in this case, 78/6 = 13.00) and subtract the mean from each score, as shown in Example 1. Then square the deviations and sum the squares, as indicated by the symbol Σ. Then enter this value in the formula along with the number of cases (N) and perform the calculations as indicated below.

Example 1:

Scores (X)	Deviations ($X - M$)	Deviations Squared (x^2)
10	10 - 13.00 = -3	9.00
11	11 - 13.00 = -2	4.00
11	11 - 13.00 = -2	4.00
13	13 - 13.00 = 0	0.00
14	14 - 13.00 = 1	1.00
19	19 - 13.00 = 6	36.00
		Σx^2 = 54.00

Thus,

$$S = \sqrt{\frac{54}{6}} = \sqrt{9.00} = 3.00$$

Of course, if you have a large number of scores, a computer is recommended for calculations.

Considering how the standard deviation is calculated should give you a feeling for the meaning of the standard deviation. As the formula indicates, it is the *square root of the squared deviations from the mean.* Thus, the larger the deviations from the mean, the larger the standard deviation. Conversely, the smaller the deviations from the mean, the smaller the standard deviation.

Appendix B

Notes on Interpreting Pearson *r* and Linear Regression

This appendix describes why a Pearson *r* can be misleadingly low and introduces a statistical procedure for making predictions for individuals when there is a reasonably strong relationship as indicated by a Pearson *r*.

Why Pearson r's Can Be Misleadingly Low

The value of a Pearson *r* can be misleadingly low for two reasons. First, its value is diminished if the variability in a group is artificially low. For the sake of illustration, let's assume that we wanted to study the relationship between height and weight in the adult population but, foolishly, selected only subjects who were exactly six feet tall. When weighing them, we found some variation in their weights. What is the correlation between height and weight among such a group? Even though there is a positive relationship between the two variables in the general adult population, the correlation in this odd sample is zero. This must be the result because those who weigh more and those who weigh less all are of the same height. (This sample cannot show that those who are taller tend to weigh more because all subjects are of the same height. Thus, the value of the Pearson *r* will equal 0.00.) A more realistic example is the relationship between scores on a college admissions test and grades earned in college. Although the test is given to all applicants in order to make admissions decisions about all of them, grades are available only for those who were admitted, and the correlation between scores and grades can be computed only for those subjects on whom we have complete data. Those for whom we have complete data are those with higher and less variable scores. Thus, the value of the Pearson *r* will be lower than would be obtained if we correlated using scores and grades for *all* applicants.

Second, an *r* can be misleadingly low if the underlying relationship is curvilinear. For example, the relationship between test-taking anxiety and performance on standardized tests might be curvilinear. That is, small amounts of anxiety might be beneficial in motivating subjects to do well on a test, but larger amounts might be detrimental. Thus, as anxiety increases, up to a point there is a positive relationship with anxiety; after reaching a critical point, as anxiety increases, there is a negative relationship. If the Pearson r is computed for such data, the negative part of the relationship will cancel out the positive part, yielding an r of near zero. Pearson recognized this problem and warned against using his statistic for describing curvilinear relationships. Other techniques such as the *correlation ratio*, which are beyond the scope of this book, are available for describing curvilinear relationships. Fortunately, such relationships are relatively rare in the social and behavioral sciences.

Making Specific Predictions for Individuals

Suppose you found a reasonably high Pearson r such as $r = .60$ between scores on a college admissions test and grades earned in a college.[1] This tells you that the admissions test is a reasonably valid predictor of grades. This does not tell you, however, how to make specific predictions for individuals who might apply to the college in the future. For example, if Marilyn has an admissions test score of 600, the Pearson r of .60 does not tell us what specific grade-point average to predict that Marilyn will earn. *Linear regression* is a statistical technique that enables us to make such predictions under most circumstances. It is beyond the scope of this book to describe this procedure, but students who have mastered correlational concepts may choose to pursue it in other books, including *Success at Statistics*, which is available from Pyrczak Publishing.

---

[1] Pearson r's for relationships between admissions test scores and grades earned in college rarely exceed .60 and often are substantially lower.

Appendix C

Table of Random Numbers

2	1	0	4	9	8	0	8	8	8	0	6	9	2	4	8	2	6
0	7	3	0	2	9	4	8	2	7	8	9	8	9	2	9	7	1
4	4	9	0	0	2	8	6	2	6	7	7	7	3	1	2	5	1
7	3	2	1	1	2	0	7	7	6	0	3	8	3	4	7	8	1
3	3	2	5	8	3	1	7	0	1	4	0	7	8	9	3	7	7
6	1	2	0	5	7	2	4	4	0	0	6	3	0	2	8	0	7
7	0	9	3	3	3	7	4	0	4	8	8	9	3	5	8	0	5
7	5	1	9	0	9	1	5	2	6	5	0	9	0	3	5	8	8
3	5	6	9	6	5	0	1	9	4	6	6	7	5	6	8	3	1
8	5	0	3	9	4	3	4	0	6	5	1	7	4	4	6	2	7
0	5	9	6	8	7	4	8	1	5	5	0	5	1	7	1	5	8
7	6	2	2	6	9	6	1	9	7	1	1	4	7	1	6	2	0
3	8	4	7	8	9	8	2	2	1	6	3	8	7	0	4	6	1
1	9	1	8	4	5	6	1	8	1	2	4	4	4	2	7	3	4
1	5	3	6	7	6	1	8	4	3	1	8	8	7	7	6	0	4
0	5	5	3	6	0	7	1	3	8	1	4	6	7	0	4	3	5
2	2	3	8	6	0	9	1	9	0	4	4	7	6	8	1	5	1
2	3	3	2	5	5	7	6	9	4	9	7	1	3	7	9	3	8
8	5	5	0	5	3	7	8	5	4	5	1	6	0	4	8	9	1
0	6	1	1	3	4	8	6	4	3	2	9	4	3	8	7	4	1
9	1	1	8	2	9	0	6	9	6	9	4	2	9	9	0	6	0
3	7	8	0	6	3	7	1	2	6	5	2	7	6	5	6	5	1
5	3	0	5	1	2	1	0	9	1	3	7	5	6	1	2	5	0
7	2	4	8	6	7	9	3	8	7	6	0	9	1	6	5	7	8
0	9	1	6	7	0	3	8	0	9	1	5	4	2	3	2	4	5
3	8	1	4	3	7	9	2	4	5	1	2	8	7	7	4	1	3

Appendix D

More About the Null Hypothesis

Many students find the null hypothesis difficult to master. Thus, the material in Sections 17 and 18 was deliberately simplified. After you have mastered these sections, you should consider the following.

Suppose you have a *directional research hypothesis* (H$_1$) that states that the population mean for Group 1 (μ_1) is higher than the population mean for Group 2 (μ_2). In other words, if the population mean for Group 2 is subtracted from the population mean for Group 1, the difference is greater than zero. Using symbols, this is how it's stated:

$$H_1: \mu_1 - \mu_2 > 0$$

The *null hypothesis* (H$_0$) that complements the research hypothesis states that if the population mean for Group 2 is subtracted from the population mean for Group 1, the difference is either zero or less than zero. Using symbols, here's how it's stated:

$$H_0: \mu_1 - \mu_2 \leq 0$$

For reasons beyond the scope of this book, significance tests of this null hypothesis are more liberal than tests of the null hypothesis described in Section 17. That is, tests of this null hypothesis are more likely to lead to rejection of the null hypothesis. Such tests are called *one-tailed tests*.

Comprehensive Review Questions

Section 1

1. The *empirical* approach to knowledge is based on
   A. deduction.
   B. reliance on authority.
   C. observation.

2. "Everyday observation is an example of the empirical approach to knowledge." This statement is
   A. true. B. false.

3. If there are 800 teachers in a school district and 100 are selected for observation, the 100 are known as a
   A. population.
   B. sample.

4. "Flawed research can be as misleading as everyday observations." This statement is
   A. true. B. false.

5. What is a primary function of statistical analysis?
   A. Planning when observations will be made.
   B. Organizing and summarizing data.
   C. Identifying a population.

Section 2

6. Treatments are given in which type of study?
   A. Experimental studies.
   B. Descriptive studies.

7. Treatments constitute which type of variable?
   A. Independent variable.
   B. Dependent variable.

8. Suppose we treated students with two types of rewards to see which one was more effective in promoting spelling achievement. *Spelling achievement* is the
   A. independent variable.
   B. dependent variable.

9. Scientists try to change the subjects in which type of study?
   A. Experimental studies.
   B. Descriptive studies.

10. A survey is an example of
    A. an experimental study.
    B. a descriptive study.

Section 3

11. *Parameters* are based on a study of a
    A. sample.
    B. population.

12. "Using volunteers when sampling is presumed to create a bias." This statement is
    A. true. B. false.

13. What is the most important characteristic of a good sample?
    A. Being free from bias.
    B. Being large.

14. "Random sampling creates sampling errors." This statement is
    A. true. B. false.

15. Using random sampling identifies
    A. an accidental sample.
    B. a sample of convenience.
    C. an unbiased sample.

Section 4

16. If subjects name their county of residence, the resulting data are at what level?
    A. Ordinal.
    B. Interval.
    C. Ratio.
    D. Nominal.

17. If a teacher ranks students from low to high on their volleyball skills, he or she is measuring at what level?
    A. Ordinal.
    B. Interval.
    C. Ratio.
    D. Nominal.

18. Which two scales of measurement tell us by *how much* subjects differ from each other?
    A. Ordinal and nominal.
    B. Interval and ordinal.
    C. Ratio and interval.
    D. Nominal and interval.

19. "The ordinal scale is a higher level of measurement than the interval scale." This statement is
    A. true. B. false.

20. "Measuring height using a tape measure is an example of the ratio scale of measurement." This statement is
    A. true. B. false.

Section 5

21. Which of the following is used to summarize data?
    A. Inferential statistics.
    B. Descriptive statistics.

22. A margin of error is an example of
    A. an inferential statistic.
    B. a descriptive statistic.

23. "It is necessary to use inferential statistics when conducting a census." This statement is
    A. true.   B. false.

24. "All populations are large." This statement is
    A. true.   B. false.

25. Which type of statistics tells us how much confidence we can have when we generalize from samples to populations?
    A. Inferential.
    B. Descriptive.

Section 6

26. "In descriptive statistics, the lowercase letter *f* stands for *function*." This statement is
    A. true.   B. false.

27. If there are 1,000 adult citizens in a town and 53% favor capital punishment, how many favor it?
    A. 53
    B. 530
    C. Some other number.

28. If 30 out of 100 parents favor school uniforms, what percentage favors them?
    A. 30%
    B. 60%
    C. 90%
    D. Some other percentage.

29. "For a proportion of .22, the corresponding percentage is 2.2%." This statement is
    A. true.   B. false.

30. "When reporting percentages, the underlying frequencies should also be reported." This statement is
    A. true.   B. false.

Section 7

31. "A frequency polygon is a drawing that shows how many subjects have each score." This statement is
    A. true.   B. false.

32. "A normal curve is also called a skewed curve." This statement is
    A. true.   B. false.

33. When a curve has a tail to the left but no tail to the right, it is said to have a
    A. positive skew.
    B. negative skew.

34. "Income in large populations is usually skewed to the right." This statement is
    A. true.   B. false.

35. "Another name for a skewed curve is 'bell-shaped curve.'" This statement is
    A. true.   B. false.

Section 8

36. "The uppercase letter X without a bar over it is a symbol for the mean." This statement is
    A. true.   B. false.

37. "The mean is the most popular average." This statement is
    A. true.   B. false.

38. "In a set of scores, the deviations from the mean have a sum of zero." This statement is
    A. true.   B. false.

39. The mean is associated with which scales of measurement?
    A. Ordinal and nominal.
    B. Interval and ordinal.
    C. Nominal and interval.
    D. Ratio and interval.

40. "The mean is an especially good average for describing skewed distributions." This statement is
    A. true.   B. false.

Section 9

41. Which average is defined as the *most frequently occurring score*?
    A. Mean.
    B. Median.
    C. Mode.

42. If the median for a set of scores equals 75, what percentage of the scores are below 75?
    A. 25%
    B. 50%
    C. 100%
    D. Some other percentage.

43. "The mean is insensitive to extreme scores." This statement is
    A. true.   B. false.

44. "In a distribution with a negative skew, the median has a higher value than the mean." This statement is
   A. true.   B. false.

45. What is the mode of the following scores?
   Scores: 1, 2, 3, 6, 6, 6
   A. 3
   B. 4
   C. 6
   D. Some other value.

Section 10

46. "A synonym for the term *variability* is *dispersion*." This statement is
   A. true.   B. false.

47. "The *range* is a statistic that describes central tendency." This statement is
   A. true.   B. false.

48. Scores that lie far outside the range of the vast majority of scores are known as
   A. *IQR*s.
   B. outliers.
   C. median points.

49. "The interquartile range is defined as the range of the middle 50% of the subjects." This statement is
   A. true.   B. false.

50. "The interquartile range is seriously affected by outliers." This statement is
   A. true.   B. false.

Section 11

51. "The standard deviation is a popular measure of variability." This statement is
   A. true.   B. false.

52. Which group has a larger standard deviation?
   A. Group X's scores: 0, 5, 10, 15, 20
   B. Group Y's scores: 1, 2, 3, 4, 5

53. "If all subjects have the same score, the value of the standard deviation is 1.00." This statement is
   A. true.   B. false.

54. In a normal distribution, what percentage of the cases lies between the mean and one standard deviation unit above the mean?
   A. 34%
   B. 50%
   C. 68%

55. In a normal distribution with a mean of 50.00 and a standard deviation of 8.00, what percentage of the cases lies between scores of 42 and 58?
   A. 34%
   B. 50%
   C. 68%

Section 12

56. For the scores on Test X and Test Y shown below, there is
   A. a direct relationship.
   B. an inverse relationship.
   C. no relationship.

Student	Test X	Test Y
Janice	25	9
Brittany	30	7
Ramon	35	4
Wallace	40	1

57. For the scores on Test D and Test E shown below, there is
   A. a direct relationship.
   B. an inverse relationship.
   C. no relationship.

Student	Test D	Test E
Buddy	303	20
Turner	343	53
Kathy	479	70
Suzanne	599	88

58. "A *direct* relationship is sometimes called a *positive* relationship." This statement is
   A. true.   B. false.

59. "In an inverse relationship, those who are high on one variable tend to be low on the other." This statement is
   A. true.   B. false.

60. "*Correlation* is considered by experts to be an excellent way to examine cause-and-effect." This statement is
   A. true.   B. false.

Section 13

61. When there is a perfect inverse relationship, what is the value of *r*?
   A. 1.00
   B. 0.00
   C. −1.00
   D. Some other value.

62. "It is possible for a relationship to be both inverse and strong." This statement is
   A. true.   B. false.

63. Which of the following values of *r* represents the strongest relationship?
    A. .64
    B. –.79
    C. 0.00

64. "An *r* of .60 is equivalent to 60%." This statement is
    A. true.   B. false.

65. "An *r* of –.95 represents a stronger relationship than an *r* of .88." This statement is
    A. true.   B. false.

Section 14

66. "The symbol for the coefficient of determination is r^2." This statement is
    A. true.   B. false.

67. If the Pearson *r* equals .30, the coefficient of determination is calculated by
    A. taking the square root of .30.
    B. multiplying .30 by .30.

68. "For an *r* of .80, the ability to predict is 64% better than zero." This statement is
    A. true.   B. false.

69. When a coefficient of determination equals .20, what percentage of the variance on one variable is *not* predicted by the other variable?
    A. 4%
    B. 20%
    C. 96%
    D. Some other percentage.

70. "When *r* = .40, the percentage of variance accounted for is 16%." This statement is
    A. true.   B. false.

Section 15

71. Putting the names of girls in one hat and boys in another hat, and drawing out 20% of the girls and 20% of the boys separately from each hat constitutes
    A. cluster sampling.
    B. stratified random sampling.
    C. simple random sampling.

72. Which of the following usually creates less sampling error?
    A. Simple random sampling.
    B. Stratified random sampling.

73. Suppose there are *500 people* in a population and you want to draw a sample using a table of random numbers. Which of the following would be an appropriate number name for the first person to whom you assign a number before using the table?
    A. 00
    B. 01
    C. 05
    D. 001

74. Which of the following will produce a greater reduction in sampling errors?
    A. Increasing the size of a sample from 800 to 900.
    B. Increasing the size of a sample from 200 to 300.

75. "It is usually better to use a small unbiased sample than a large biased one." This statement is
    A. true.   B. false.

Section 16

76. "Using random sampling guarantees freedom from sampling errors." This statement is
    A. true.   B. false.

77. "According to the central limit theorem, the sampling distribution of means is skewed." This statement is
    A. true.   B. false.

78. The larger the variability in a population, the
    A. larger the standard error of the mean.
    B. smaller the standard error of the mean.

79. If *m* = 40.00 and SE_M = 3.00, what are the limits of the 68% confidence interval for the mean?
    A. 37.00 and 40.00
    B. 40.00 and 43.00
    C. 37.00 and 43.00
    D. Some other values.

80. If you increase the sample size, what effect does this have on the size of the standard error of the mean?
    A. It increases it.
    B. It decreases it.

Section 17

81. Which of the following is a correct statement of the null hypothesis?
    A. There is a true difference between the means.
    B. There is no true difference between the means.

82. Which of the following is a symbol for the null hypothesis?
   A. H_0
   B. H_1

83. "For a given study, the research hypothesis and the null hypothesis usually say the same thing." This statement is
   A. true.   B. false.

84. Which type of hypothesis predicts that one particular group's mean will be higher than another group's mean?
   A. Directional hypothesis.
   B. Nondirectional hypothesis.
   C. Null hypothesis.

85. The null hypothesis states that the difference between the means in a population
   A. equals zero.
   B. is greater than zero.
   C. is less than zero.

Section 18

86. Which of the following yields a probability?
   A. A descriptive statistic.
   B. An inferential test.

87. At what point is it conventional to regard something as being untrue?
   A. When the probability is less than .05.
   B. When the probability is greater than .05.
   C. When the probability is greater than .10.
   D. When the probability is less than .50.

88. The null hypothesis can be rejected with the greatest confidence when which one of the following is true?
   A. $p < .05$
   B. $p < .01$
   C. $p < .001$

89. Rejecting the null hypothesis when, in reality, it is true is known as a
   A. Type I error.
   B. Type II error.

90. By conventional standards, if $p < .01$, we would declare the difference to be statistically
   A. insignificant.
   B. significant.

Section 19

91. "A t test yields a probability." This statement is
   A. true.   B. false.

92. "The smaller the sample, the more likely the null hypothesis will be rejected." This statement is
   A. true.   B. false.

93. Under which of the following circumstances is the null hypothesis more likely to be rejected?
   A. When there is a small observed difference between means.
   B. When there is a large observed difference between means.

94. "Dependent data may have less sampling error than independent data." This statement is
   A. true.   B. false.

95. If subjects are matched (i.e., paired) across experimental and control groups, the resulting data are
   A. independent.
   B. dependent.

Section 20

96. "Reporting a t test makes it unnecessary to report the values of the means and standard deviations." This statement is
   A. true.   B. false.

97. If you read that $t = 0.452$, $df = 100$, $p > .05$, what should you conclude?
   A. The difference is statistically significant.
   B. The difference is not statistically significant.

98. "If a t test yields $p < .05$, the null hypothesis normally would be rejected." This statement is
   A. true.   B. false.

99. If we say that we are rejecting the null hypothesis, what else are we concluding?
   A. The difference is statistically significant.
   B. The difference is not statistically significant.

100. "It is safe to assume that if a difference is statistically significant, it is of practical significance." This statement is
   A. true.   B. false.

Section 21

101. "ANOVA can be used to test for the difference(s) between only two means." This statement is
   A. true.   B. false.

102. "The acronym *ANOVA* stands for *Analysis of Variance*." This statement is
A. true.   B. false.

103. For the typical consumer of research, which one of the following values in an ANOVA table is of greatest interest?
A. Mean squares.
B. The value of p.
C. The value of F.

104. Suppose you read the following: $F = 0.641$, $df = 3, 29, p > .05$. What conclusion would you normally draw about the null hypothesis?
A. Reject it.
B. Do not reject it.

105. Suppose you read the following: $F = 3.50$, $df = 2, 20, p < .05$. What conclusion would you normally draw about statistical significance?
A. It is statistically significant.
B. It is not statistically significant.

Section 22

106. Suppose subjects were classified according to their religion and their country of origin in order to compare means for both religious groups and national origin groups. This would call for a
A. One-way ANOVA.
B. Two-way ANOVA.

107. In order to examine an *interaction*, you
A. temporarily ignore one way that the subjects were classified while examining the results of the other way they were classified.
B. look at both ways subjects were classified at the same time in order to see how the two classification variables affect each other.

108. "In the table below, there appears to be an interaction." This statement is
A. true.   B. false.

	X	Y
D	M = 40.00	M = 30.00
E	M = 30.00	M = 40.00

109. "In the table below, there appears to be an interaction." This statement is
A. true.   B. false.

	S	T
U	M = 300.00	M = 200.00
V	M = 350.00	M = 250.00

110. "In the table below, there appear to be two main effects." This statement is
A. true.   B. false.

	S	T
U	M = 40.00	M = 50.00
V	M = 30.00	M = 20.00

Section 23

111. "For nominal data, we normally report frequencies and percentages instead of means and standard deviations." This statement is
A. true.   B. false.

112. The symbol for chi square is
A. p
B. r^2
C. χ^2
D. p^2

113. "For the data in the following table, a two-way chi square would be an appropriate test of significance." This statement is
A. true.   B. false.

	Yes	No
Men	m = 30	m = 40
Women	m = 40	m = 30

114. Suppose you read that as the result of a chi square test, $p < .001$. By conventional standards, what decision should be made about the null hypothesis?
A. Reject it.
B. Do not reject it.

115. Suppose you read that as the result of a chi square test, $p < .05$. By conventional standards, what decision should be made about statistical significance?
A. It is significant.
B. It is not significant.